T0324905

Problem Books in Mathematics

Edited by P.R. Halmos

T. Cacoullos

Exercises in Probability

With 22 Illustrations

Springer-Verlag
New York Berlin Heidelberg
London Paris Tokyo

T. Cacoullos
Department of Mathematics
University of Athens
Athens 157 10
Greece

Editor

Paul R. Halmos
Department of Mathematics
Santa Clara University
Santa Clara, CA 95053
U.S.A.

Mathematics Subject Classification (1980): 60-01

Library of Congress Cataloging-in-Publication Data
Cacoullos, T. (Theophilos)
  Exercises in probability.
  (Problems in mathematics)
  Translated from the Greek.
  Bibliography: p.
  1. Probabilities—Problems, exercises, etc.   I. Title.
II. Series.
QA273.25.C33   1989   519.2′076   88-4932

Printed on acid-free paper.

Original Greek edition published in two volumes, *Askiseis Theorias Pithanotiton* (*Exercises in Probability*) and *Askiseis-Provlemata Theorias Pithanotition* (*Exercises-Problems in Probability Theory*), in 1971 and 1972, respectively, by Leousis-Mastroyiannis, Athens, Greece.

Typeset by Asco Trade Typesetting Ltd., Hong Kong.
Printed and bound by R.R. Donnelley & Sons, Harrisonburg, Virginia.
Printed in the United States of America.

9 8 7 6 5 4 3 2 1

ISBN 0-387-96735-4 Springer-Verlag New York Berlin Heidelberg
ISBN 3-540-96735-4 Springer-Verlag Berlin Heidelberg New York

Preface

This problem book can serve as a companion text for an introductory or intermediate-level one- or two-semester probability course, usually given to undergraduate or first-year graduate students in American universities. Those who will benefit most are those with a good grasp of calculus and the inclination and patience required by mathematical thinking; and yet many more, with less formal mathematical background and sophistication, can benefit greatly from the elements of theory and solved problems, especially of Part I. In particular, the important part of discrete probabilities with problems of a combinatorial nature, relating to finite sample spaces and equally likely cases, may be challenging and entertaining to the student and amateur alike, through the spectrum from mathematics to the social sciences.

For some time, since I have been teaching courses in probability theory both to undergraduates at the University of Athens (School of Science) and graduate students in American universities, I have noted the need of students for a systematic collection of elementary and intermediate-level exercises in this area. The relevant international bibliography is rather limited. The aim of the present collection is to fill such a gap.

This book is based on the two volumes of the Greek edition entitled *Exercises in Probability* by the same author. Of the 430 exercises contained in the original Greek edition (1972), 329 (with solutions) were selected for the present translation. A small number of problems in stochastic processes is not included here, since it is felt that a minimal treatment of this important area requires more theory and a larger number of problems. However, an addendum of over 160 exercises and certain complements of theory and problems (with answers or hints) from the recent (1985) second Greek edition is included here as Supplement I: Miscellaneous Exercises, and as Supplement II: Complements and Problems.

In addition to the Supplements, the present book is divided into three parts. Part I: Elementary Probabilities (Chapters 1–4), Part II: Advanced Topics (Chapter 5–10), and Part III: Solutions (to the exercises of Part I and Part II). In most chapters, the exercises are preceded by some basic theory and formulas.

In Part I (with 170 exercises) emphasis is given to classical probabilities (equally likely cases). It concerns mainly one-dimensional, discrete, and continuous distributions (Chapter 2). The expected value, variance, and moments of a distribution are treated in Chapter 3. Chapter 4, in spite of its elementary nature, contains quite a few challenging problems. In general, problems of a higher level of difficulty are marked with an asterisk throughout the text.

Part II deals with more advanced topics: multivariate distributions (Chapter 5), generating and characteristic functions (Chapter 6), distribution of functions of random variables (Chapter 7), and Laws of Large Numbers and Central Limit Theorems (Chapter 8). Chapter 9 deals with special topics: stochastic inequalities, geometrical probabilities, and applications of difference equations in probability. The last chapter (Chapter 10) contains general overview exercises.

Clearly, a collection of problems owes a lot to the relevant bibliography. The exercises come from many sources, some of them are new, a large number are variations of more or less standard exercises and problems of elementary or intermediate level, and a few are based on research papers. Such problems, as well as complements, are also marked with an asterisk. The level and sources are indicated in the select Bibliography.

It is hoped that the present edition will make proportionately as many friends as it did among the ten thousand or so students all over Greece who have used the text since its first edition in 1971–72. Naturally, one benefits from such a text by first really trying to solve a problem, before falling into the "temptation" of looking up the solution. That is why solutions are given separately, in the last part of the book.

Thanks are due to my colleagues Dr. Ch. Charalambides and Dr. Ourania Chrysaphinou for preparing some solutions for the first Greek edition, as well as preparing a first draft of a major part of this translation.

New York   T. CACOULLOS
June, 1987   Visiting Professor
                                                              Columbia University

Contents

Contents

PART I

ELEMENTARY PROBABILITIES

CHAPTER 1

Basic Probabilities. Discrete Spaces

Basic Definitions and Formulas

1. *Sample space* Ω: The totality of possible outcomes of a random (chance) experiment.

The individual outcomes are called *elementary* or *simple events* or *points* (*cases* in the classical definition of probability by Laplace).

2. *Discrete sample space* Ω: Ω is at most a denumerable set of points.

3. *Events*: Subsets* of Ω.

4. *Union* of n events $A_1, \ldots, A_n$, denoted by $A_1 \cup A_2 \cdots \cup A_n$, is the realization of at least one of the events $A_1, \ldots, A_n$.

5. *Realization* or *occurrence* of an event A means the appearance of an element (point) in A.

6. *Intersection or product* of two events A and B, denoted by $A \cap B$ or AB, is the simultaneous realization of both events A and B. Similarly for n events.

7. *Complement* of A, denoted by A^c or A', means A does not occur.

8. The *difference* of two events A and B is defined by $A - B = AB^c$, i.e., A occurs but not B. Thus $A^c = \Omega - A$.

9. *Sure or certain event*: The sample space Ω.

10. *Impossible event*: The complement of the sure event, i.e., the empty set.

---

* Rigorously, measurable subsets of Ω, i.e., members of a so-called Borel or σ-field $B(\Omega)$ on Ω: a family of subsets of Ω containing Ω itself and closed under the set operations of complementation and countable union.

11. *Probability.* A probability (function) P is a set function on Ω (strictly on the Borel field $B(\Omega)$ on Ω) which satisfies the following three axioms of Kolmogorov:

(I) $P(\Omega) = 1$.

(II) For every event A, $P(A) \geq 0$.

(III) For every sequence of mutually exclusive events

$$A_1, A_2, \ldots, A_n, \ldots, \qquad A_i A_j = \varnothing, \qquad i \neq j,$$

$$P(A_1 \cup A_2 \cup \cdots \cup A_n \cup \cdots) = P(A_1) + P(A_2) + \cdots + P(A_n) + \cdots.$$

12. (a) *Equally likely cases.* Let Ω be a finite sample space with N elements (points) $\omega_1, \omega_2, \ldots, \omega_N$; if

$$P[\{\omega_i\}] = \frac{1}{N}, \qquad i = 1, \ldots, N,$$

then the ω_i are called equally likely elementary events (cases).

(b) *Laplace definition of probability.* Let $A \subset \Omega$, where the points of Ω are assumed equally likely. Then

$$P[A] = \frac{\text{number of points in } A}{\text{number of points in } \Omega}$$

$$= \frac{\text{number of favorable cases to } A}{\text{number of cases in } \Omega}.$$

13. *Basic formulas for probabilities.*

(i) $P(A^c) = 1 - P(A)$.

(ii) $P(A - B) = P(A) - P(AB)$.

(iii) Addition theorem of probabilities: $P(A \cup B) = P(A) + P(B) - P(AB)$.

Poincaré's theorem. *For n events* $A_1, \ldots, A_n$,

$$P(A_1 \cup A_2 \cup \cdots \cup A_n) = S_1 - S_2 + \cdots + (-1)^{n-1} S_n, \qquad (1.1)$$

where, for each $k = 1, 2, \ldots, n$, S_k *is defined by*

$$S_k = \sum_{i_1 < i_2 < \cdots < i_k} P(A_{i_1} A_{i_2} \ldots A_{i_k}), \qquad i_1 \geq 1, \quad i_k \leq n,$$

i.e., the summation is over all combinations of the n events taken k at a time ($k = 1, \ldots, n$).

14. *Conditional probability* of B given A:

$$P(B|A) = \frac{P(AB)}{P(A)}, \qquad P(A) > 0.$$

15. *Multiplication formula* for probabilities:

$$P(AB) = P(B|A)P(A) = P(A|B)P(B).$$

In general,

$$P(A_1 A_2 \ldots A_n) = P(A_1)P(A_2|A_1)P(A_3|A_1 A_2) \ldots P(A_n|A_1 A_2 \ldots A_{n-1}). \quad (1.2)$$

16. *Independent events* (statistically or stochastically):
(a) Two events A, B: if $P(AB) = P(A)P(B)$.
(b) n events $A_1, \ldots, A_n$: if $P(A_{i_1} \ldots A_{i_k}) = P(A_{i_1}) \ldots P(A_{i_k})$ for $1 \leq i_1 < i_2 < \cdots < i_k \leq n$, $2 \leq k \leq n$.

17. *Independent experiments.* Let $E_1, E_2, \ldots, E_n$ be n chance experiments and $\Omega_1, \ldots, \Omega_n$ the corresponding sample spaces. The experiments are called *stochastically* or *statistically independent* or simply *independent* if for every $A_i \subset \Omega_i$ $(i = 1, 2, \ldots, n)$,

$$P(A_1 A_2 \ldots A_n) = P(A_1)P(A_2) \ldots P(A_n).$$

Remark. Physically independent experiments are assumed statistically independent, e.g., successive throws of a coin, a die, etc.

18. *Total probability formula.* If $B_i \cap B_j = \varnothing$, $i \neq j$ $(i, j = 1, \ldots, n)$ and $A \subset (B_1 \cup B_2 \cup \cdots \cup B_n)$, then

$$P(A) = P(A|B_1)P(B_1) + \cdots + P(A|B_n)P(B_n). \quad (1.3)$$

19. *Bayes's theorem (formula).* As in Formula 18,

$$P(B_i|A) = \frac{P(A|B_i)P(B_i)}{P(A|B_1)P(B_1) + \cdots + P(A|B_n)P(B_n)}, \qquad i = 1, \ldots, n. \quad (1.4)$$

20. *Combinatorics*
Binomial coefficient. For every real x and any positive integer k we define

$$\binom{x}{k} = \frac{x(x-1)\cdots(x-k+1)}{k!}.$$

In particular, when x is a positive integer, then

$$\binom{x}{k} = \frac{x!}{k!\,(x-k)!}$$

equals the number of combinations of x distinct objects taken k at a time. *Pascal's triangle* is based on the recursive relation

$$\binom{n+1}{k} = \binom{n}{k-1} + \binom{n}{k}.$$

21. Let S be a set (population) of N distinct elements.
(a) A subset of k elements (a k-tuple) is called a sample of size k (a *combination*).
(b) An ordered k-tuple of elements of S is called an *ordered sample* of size

k (a *permutation*). There are

$$\binom{N}{k}k! = \frac{N!}{(N-k)!} \equiv (N)_k = N(N-1)\cdots(N-k+1)$$

permutations of N objects taken k at a time.

(c) S can be divided into k subsets, the first subset (subpopulation) containing r_1 elements, the second r_2, etc., the kth r_k objects in

$$\frac{N!}{r_1!\,r_2!\dots r_k!} \equiv \binom{N}{r_1,\dots,r_k} \tag{1.5}$$

ways; (1.5) is called a *multinomial coefficient*.

22. *Placing balls into cells.*
(a) s distinguishable balls can be placed into n cells in n^s ways.
(b) s nondistinguishable balls can be placed into n cells in

$$\binom{n+s-1}{n-1} = \binom{n+s-1}{s} \tag{1.6}$$

ways; that is, also the number of ordered n-tuples $(r_1, r_2, \dots, r_n)$ of integers $r_i \geq 0$ which are solutions of the equation

$$r_1 + r_2 + \cdots + r_n = s.$$

The numbers r_i are referred to as the (cell) *occupancy numbers* (see Exercise 44).

23. *Sampling without replacement.* From an urn containing W white balls and B black balls, n balls are drawn (one after another or at once) at random. The probability p_r that r white balls are drawn is given by

$$p_r = \frac{\binom{W}{r}\binom{B}{n-r}}{\binom{N}{r}} = \frac{\binom{n}{r}\binom{N-n}{W-r}}{\binom{N}{W}}, \qquad N = W+B, \tag{1.7}$$

$$\max(0, n-B) \leq r \leq \min(n, W).$$

The probabilities p_r define the so-called *hypergeometric distribution*.

24. *Multiplication principle.* If an "operation" A_1 can be performed in n_1 ways, another "operation" A_2 in n_2 ways, etc., the kth "operation" in n_k ways, then the k "operations" can be performed, one after another, in $n_1 n_2 \dots n_k$ ways.

This is equivalent to the proposition: If A_i consists of n_i points $(i = i, \dots, k)$, then the size (number of points) of the *cartesian product*

$$A_1 \times A_2 \times \cdots \times A_k = \{(x_1, \dots, x_k): x_i \in A_i, i = 1, \dots, k\}$$

is equal to $n_1 n_2 \dots n_k$.

25. *The principle of mathematical induction.* For every integer n let P_n be a proposition which is either true or false, i.e., P_n may be true for some n and false for other values of n; if, (i) P_1 is true, (ii) for every n, P_n true $\Rightarrow P_{n+1}$ true, then P_n is true for every n.

Exercises

1. Sets. Events

1. Let Ω denote the totality of students at the University of Athens and A_1, A_2, A_3, A_4, the sets of freshmen, sophomores, juniors, and seniors, respectively. Moreover, let F denote the set of female students and C the set of Cypriot students. Express in words each of the following sets:
(a) $(A_1 \cup A_2) \cap F$; (b) FC'; (c) $A_1 F'C$; (d) $A_3 FC'$; (e) $(A_1 \cup A_2)CF$.

2. Give the simplified forms of the sets:
(a) $(A \cup B) \cap (A \cup C)$; (b) $(A \cup B) \cap (A' \cup B)$; (c) $(A \cup B) \cap (A' \cup B) \cap (A' \cup B')$.

3. Express each of the following events in terms of the events A, B, and C, and the operations of complementation, union, and intersection:
(a) at least one of the events A, B, C occurs;
(b) at most one of the events A, B, C occurs;
(c) none of the events A, B, C occurs;
(d) all three events occur;
(e) exactly one of the events A, B, C occurs;
(f) A and B occur but not C;
(g) A occurs, if not then B does not occur either.
Give the Venn diagram for each of the above.

4. Let Ω denote the sample space corresponding to the chance experiment of tossing a coin three times. Let A be the event that heads appear exactly twice, let B be the event that at least two heads appear, and let C the event that heads appear when tails have appeared at least once.
(a) Give the elements of A, B, C;
(b) Describe the events: (i) $A'B$; (ii) $A'B'$; (iii) AC.

In each of the Exercises 5–8, give the number of points in the sample space Ω and the points of the events defined therein.

5. A family has 4 children. Let the events: A: boys and girls alternate, B: the first and fourth child are boys, C: as many boys as girls, and D: three successive children of the same sex.

6. A salesman is arranging his schedule for visiting each of three towns a, b, c, twice. A: the first and last visit are in a.

7. Certain diseases cause cell disorders which can be distinguished into four categories. Blood tests are conducted on each of 4 patients and in each case the category is noted. Let A: the patients belong to the same category, B: two patients belong to the same category.

8. An elevator carries two persons and stops at three floors. Let A: they get off at different floors, B: one gets off at the first floor.

2. Combinatorics

9. In how many ways can a lady having 10 dresses, 5 pairs of shoes, and 2 hats be dressed?

10. In how many ways can we place in a bookcase two works each of three volumes and two works each of four volumes, so that the volumes of the same work are not separated?

11. In how many ways can r objects be distributed to n persons if there is no restriction on the number of objects that a person may receive?

12. In how many ways can 5 boys and 5 girls be seated around a table so that no 2 boys sit next to each other?

13. Eight points are chosen on the circumference of a circle. How many chords can be drawn by joining these points in all possible ways? If the 8 points are considered vertices of a polygon, how many triangles and how many hexagons can be formed?

14. In how many ways can n persons be seated around a table if 2 arrangements are regarded as the same when each person has the same right and left neighbors? At a dinner at which the n persons are seated randomly, what is the probability that a specific husband is seated next to his wife?

15. In how many ways can 20 recruits be distributed into 4 groups each consisting of 5 recruits? In how many ways can they be distributed into 4 camps, each camp receiving 5 recruits?

16. Using 7 consonants and 5 vowels, how many words consisting of 4 consonants and 3 vowels can we form?

3. Properties of Binomial Coefficients

17. Show that for every x and for every positive integer n:

(i) $\dbinom{x}{n-1} + \dbinom{x}{n} = \dbinom{x+1}{n}$;

(ii) $\dbinom{x}{n} = \dbinom{x-1}{n-1} + \dbinom{x-2}{n-1} + \cdots + \dbinom{n-1}{n-1}$.

18. If n, m, r, are positive integers prove that
$$\binom{m}{0}\binom{n-m}{r} + \binom{m}{1}\binom{n-m}{r-1} + \cdots + \binom{m}{r}\binom{n-m}{0} = \binom{n}{r}.$$

19. For every integer $n \geq 2$, show the following:

(i) $1 - \binom{n}{1} + \binom{n}{2} + \cdots + (-1)^n\binom{n}{n} = 0$;

(ii) $\binom{n}{1} + 2\binom{n}{2} + 3\binom{n}{3} + \cdots = n2^{n-1}$;

(iii) $2 \cdot 1 \cdot \binom{n}{2} + 3 \cdot 2 \cdot \binom{n}{3} + 4 \cdot 3 \cdot \binom{n}{4} + \cdots = n(n-1)2^{n-2}$.

20. Using Pascal's triangle, show that for every a and positive integers r, n:

(i) $\sum_{k=0}^{n}\binom{a-k}{r} = \binom{a+1}{r+1} - \binom{a-n}{r+1}$;

(ii) $\sum_{k=0}^{n}(-1)^k\binom{a}{k} = (-1)^n\binom{a-1}{n}$.

21. Using the result of Exercise 18, show that
$$\binom{n}{0}^2 + \binom{n}{1}^2 + \cdots + \binom{n}{n}^2 = \binom{2n}{n}.$$

22. Using the result of Exercise 21, prove that
$$\sum_{v=0}^{n}\frac{(2n)!}{(v!)^2[(n-v)!]^2} = \binom{2n}{n}^2.$$

23. Using the binomial theorem, show by induction that
$$\sum_{k=1}^{n}\binom{n}{k}\frac{(-1)^{k-1}}{k} = \sum_{k=1}^{n}\frac{1}{k}.$$

24. As in Exercise 23, show that for positive integers r and n
$$\sum_{k=0}^{n}\binom{k+r-1}{r-1} = \binom{n+r}{r}.$$

Show that this is a special case of Problem 20(ii).

25. In the expansion of $(1 + x)^n$ where $x > 0$ and n is a positive integer, let a_k be the term containing x^k. Find the values of k for which a_k becomes maximum. What is the maximum term of $(1 + e)^{100}$?

4. Properties of Probability

26. Given $P(A) = 1/3$, $P(B) = 1/4$, $P(AB) = 1/6$, find the following probabilities:
$$P(A'), \quad P(A' \cup B), \quad P(A \cup B'), \quad P(AB'), \quad P(A' \cup B').$$

27. Given $P(A) = 3/4$ and $P(B) = 3/8$, show that:
(a) $P(A \cup B) \geq 3/4$,
(b) $1/8 \leq P(AB) \leq 3/8$. Give inequalities analogous to (a) and (b) for $P(A) = 1/3$ and $P(B) = 1/4$.

28. For any two events A and B show that:

$$P(AB) - P(A)P(B) = P(A')P(B) - P(A'B) = P(A)P(B') - P(AB').$$

29. By induction show that for arbitrary events $A_1, A_2, \ldots, A_n$, the following inequality (due to Bonferoni) holds:

$$P(A_1 A_2 \ldots A_n) \geq \sum_{i=1}^{n} P(A_i) - (n - 1) = 1 - \sum_{i=1}^{n} P(A_i^c).$$

30. For any three events A, B, and C, show that (cf. (1.1))

$$P[A \cup B \cup C] = P(A) + P(B) + P(C) - P(AB) - P(AC)$$
$$- P(BC) + P(ABC).$$

Application: The percentages of students who passed courses A, B, and C are as follows: A: 50%, B: 40%, C: 30%, A and B: 35%, B and C: 20%, and 15% passed all three courses. What is the percentage of students who succeeded in at least one of the three courses?

31. A box contains balls numbered $1, 2, \ldots, n$. A ball is drawn at random:
(a) What is the probability that its label number is divisible by 3 or 4?
(b) Examine the case in (a) as $n \to \infty$.

32. In terms of $P(A)$, $P(B)$, $P(C)$, $P(AB)$, $P(AC)$, $P(BC)$, and $P(ABC)$ express, for $k = 0, 1, 2, 3$, the probabilities that:
(i) exactly k of the events A, B, and C, occur;
(ii) at least k of the events A, B, C occur.

33. By induction prove Poincaré's theorem, i.e., formula (1.1).

34.* If $d(A, B) = P(A \triangle B)$, show that d has all the properties of a distance function; the symmetric difference $A \triangle B$ between two sets A and B is defined by $A \triangle B = AB' \cup A'B$.

5. Classical Probabilities. Equally Likely Cases

35. Three winning tickets are drawn from an urn of 100 tickets. What is the probability of winning for a person who buys: (a) 4 tickets? (b) only one ticket?

36. A bakery makes 80 loaves of bread daily. Ten of them are underweight. An inspector weighs 5 loaves at random. What is the probability that an underweight loaf will be discovered?

37. Find the probability that among seven persons:
(a) no two were born on the same day of the week (Sunday, Monday, etc.);
(b) at least two were born on the same day;
(c) two were born on a Sunday and two on a Tuesday.

38. A group of $2N$ boys and $2N$ girls is randomly divided into two equal groups. What is the probability that each group has the same number of boys and girls?

39. The coefficients a, b, c of the quadradic equation $ax^2 + bx + c = 0$ are determined by throwing a die three times. Find the probabilities that: (a) the roots are real; (b) the roots are complex.

40. In the game of poker a "hand of cards" means 5 cards randomly selected (without replacement) from a deck of 52 cards. What is the probability that a hand of cards:
(a) consists of an ace, a queen, a jack, a king, and a ten of the same suit?
(b) contains 4 cards of the same denomination (aces, etc.)?
(c) consists of cards with consecutive values, except aces, jacks, kings, and queens?

41. In bridge, "a hand of cards" consists of 13 cards drawn at random (without replacement) from a deck of 52 cards (the 52 cards are equally distributed to four players). Find the probabilities that "a hand of cards" will contain:
(i) v_1 clubs, v_2 spades, and v_3 diamonds;
(ii) v aces ($v = 0, 1, \ldots, 4$);
(iii) v_1 aces and v_2 kings.
In a bridge game find the probabilities that:
(a) each player has an ace;
(b) some player has all the aces;
(c) some player has v_1 aces and his partner v_2 aces ($v_1 + v_2 \le 4$).

42. Six girls are to enter a dance with 10 boys to form a ring so that every girl is between two boys:
(a) What is the probability that some specified boy remains between 2 boys?
(b) A spectator notices that a certain girl enters next to a certain boy. Is it random?

43. Five letters are selected at random one after another from the 26 letters of the English alphabet; (a) with replacement, (b) without replacement. Find, for each of the cases (a) and (b), the probabilities that the word formed: (i) contains an "a", (ii) consits of vowels, (iii) is the word "woman".

44.* Prove that the number of ways in which r indistinguishable balls can be distributed in n cells (or, equivalently, the number of different solutions of the equation $x_1 + x_2 + \cdots + x_n = r$ where $x_i \geq 0 \, (i = 1, 2, \ldots, n)$ are integers), equals $\binom{n + r - 1}{n - 1} = \binom{n + r - 1}{r}$.

45.* Consider r indistinguishable balls randomly distributed in n cells (Bose–Einstein statistics). What is the probability that exactly m cells remain empty?

46. From an ordinary deck of 52 cards, cards are drawn successively until an ace appears. What is the probability that the first ace will appear: (a) at the nth draw? (b) after the nth card?

47.* *Birthday problem.* In a classroom there are v students.

(a) What is the probability that at least two students have the same birthday?

(b) What is the minimum value of v which secures probability $1/2$ that at least two have a common birthday?

48.* N letters are placed at random in N envelopes. Show that the probability that each letter will be placed in a wrong envelope is $\sum\limits_{k=2}^{N} (-1)^k \left(\dfrac{1}{k!}\right)$.

49.* An urn contains nr balls numbered $1, 2, \ldots, n$ in such a way that r balls bear the same number i for each $i = 1, 2, \ldots, n$. N balls are drawn at random without replacement. Find the probability that, (a) exactly m of the numbers will appear in the sample, (b) each of the n numbers will appear at least once.

50.* (*Continuation*). Balls are drawn until each of the numbers $1, 2, \ldots, n$ appears at least once. What is the probability that m balls will be needed?

51. N men run out of a men's club after a fire and each takes a coat and a hat. Prove that:

(a) the probability that no one will take his own coat and hat is

$$\sum_{k=1}^{N} (-1)^k \frac{(N - k)!}{N! \, k!};$$

(b) the probability that each man takes a wrong coat and a wrong hat is

$$\left[\sum_{k=2}^{N} (-1)^k \frac{1}{k!} \right]^2.$$

52.* Suppose every packet of the detergent TIDE contains a coupon bearing one of the letters of the word TIDE. A customer who has all the letters of the word gets a free packet. All the letters have the same possibility of appearing in a packet. Find the probability that a housewife who buys 8 packets will get: (i) one free packet, (ii) two free packets.

53.* An urn contains N_1 white and N_2 black balls. When two balls are randomly drawn the probability that both be white is $1/2$.

(a) What is the minimum value of N_1?

(b) What is the minimum value of N_1 when N_2 is an even number?

(c) What is the minimum value of the total number $N = N_1 + N_2$ of balls in the urn?

6. Independent Events. Conditional Probability

54. If the events A, B are independent, show that the pairs (A, B'), (A', B), (A', B') also consist of independent events.

55. If the events A, B, and C are mutually independent, then the pairs (A, BC), (B, AC), (C, AB) also consist of independent events.

56. Suppose that for the independent events A, B, and C we have $P(A) = a$, $P(A \cup B \cup C) = 1 - b$, $P(ABC) = 1 - c$, and $P(A'B'C) = x$. Prove that the probability x satisfies the equation

$$ax^2 + [ab - (1 - a)(a - c - 1)]x + b(1 - a)(1 - c) = 0.$$

Hence conclude that

$$c > \frac{(1 - a)^2 + ab}{1 - a}.$$

Moreover, show that

$$P(B) = \frac{(1 - c)(x + b)}{ax}, \qquad P(C) = \frac{x}{x + b}.$$

57. Let the events A_1, A_2, ..., A_n be independent and $P(A_i) = p$ $(i = 1, 2, ..., n)$. What is the probability that:

(a) at least one of the events will occur?

(b) at least m of the events will occur?

(c) exactly m of the events will occur?

58. A die is thrown as long as necessary for an ace or a 6 to turn up. Given that no ace turned up at the first two throws, what is the probability that at least three throws will be necessary?

59. A parent particle can be divided into 0, 1, or 2 particles with probabilities 1/4, 1/2, 1/4, respectively. It disappears after splitting. Beginning with one particle, the progenitor, let us denote by X_i the number of particles in the i generation. Find, (a) $P(X_2 > 0)$, (b) the probability that $X_1 = 2$ given that $X_2 = 1$.

60. An urn contains n balls numbered 1, 2, ..., n. We select at random r balls, (a) with replacement, (b) without replacement. What is the probability that the largest selected number is m?

61. Two athletic teams A and B play a series of independent games until one of them wins 4 games. The probability of each team winning in each game equals $1/2$. Find the probability that the series will end, (a) in at most 6 games, (b) in 6 games given that team A won the first two games.

62. It is suspected that a patient has one of the diseases A_1, A_2, A_3. Suppose that the population percentages suffering from these illnesses are in the ratio $2:1:1$. The patient is given a test which turns out to be positive in 25% of the cases of A_1, 50% of A_2, and 90% of A_3. Given that out of three tests taken by the patient two were positive, find the probability for each of the three illnesses.

63. The population of Nicosia (Cyprus) is 75% Greek and 25% Turkish. 20% of the Greeks and 10% of the Turks speak English. A visitor to the town meets someone who speaks English. What is the probability that he is a Greek? Interpret your answer in terms of the population of the town.

64. Two absent-minded room mates, mathematicians, forget their umbrellas in some way or another. A always takes his umbrella when he goes out, while B forgets to take his umbrella with probability $1/2$. Each of them foregets his umbrella at a shop with probability $1/4$. After visiting three shops they return home. Find the probability that:
  (a) they have both umbrellas;
  (b) they have only one umbrella;
  (c) B has lost his umbrella given that there is only one umbrella after their return.

65. Consider families of n children and let A be the event that a family has children of both sexes, and let B be the event that there is at most one girl in the family. Show that the only value of n for which the events A and B are independent is $n = 3$, assuming that each child has probability $1/2$ of being a boy.

66. At the college entrance examination each candidate is admitted or rejected according to whether he has passed or failed the test. Of the candidates who are really capable, 80% pass the test; and of the incapable, 25% pass the test. Given that 40% of the candidates are really capable, find the proportion of capable college students.

67. *Huyghens problem.* A and B throw alternately a pair of dice in that order. A wins if he scores 6 points before B gets 7 points, in which case B wins. If A starts the game what is his probability of winning?

68. Three players P_1, P_2, and P_3 throw a die in that order and, by the rules of the game, the first one to obtain an ace will be the winner. Find their probabilities of winning.

69. N players $A_1, A_2, \ldots, A_N$ throw a biased coin whose probability of heads equals p. A_1 starts (the game), A_2 plays second, etc. The first one to throw heads wins. Find the probability that A_k ($k = 1, 2, \ldots, N$) will be the winner.

70. Urn A contains w_1 white balls and b_1 black balls. Urn B contains w_2 white balls and b_2 black balls. A ball is drawn from A and is placed into B, and then a ball is transferred from B to A. Finally, a ball is selected from A. What is the probability that the ball will be white?

71. Ten percent of a certain population suffer from a serious disease. A person suspected of the disease is given two independent tests. Each test makes a correct diagnosis 90% of the time. Find the probability that the person really has the illness:
(a) given that both tests are positive;
(b) given that only one test is positive.

72. A player randomly chooses one of the coins A, B. Coin A has probability of heads 3/4, and coin B has probability of heads 1/4. He tosses the coin twice.
(a) Find the probability that he obtains: (i) two heads, (ii) heads only once.
(b) Instead of the above procedure, suppose the player can choose an unbiased (symmetric) coin, which he tosses twice. Which procedure should he follow in order to maximize the probability of at least one head?

73. Urn A contains 5 black balls and 6 white balls, and urn B contains 8 black balls and 4 white balls. Two balls are transferred from B to A and then a ball is drawn from A.
(a) What is the probability that this ball is white?
(b) Given that the ball drawn is white, what is the probability that at least one white ball was transferred to A?

74. The Pap test makes a correct diagnosis with probability 95%. Given that the test is positive for a lady, what is the probability that she really has the disease? How do you interpret this? Assume that one in every 2,000 women, on average, has the disease.

75. A secretary goes to work following one of three routes A, B, C. Her choice of route is independent of the weather. If it rains, the probabilities of arriving late, following A, B, C, are 0.06, 0.15, 0.12, respectively. The corresponding probabilities, if it does not rain, are 0.05, 0.10, 0.15.
(a) Given that on a sunny day she arrives late, what is the probability that she took route C? Assume that, on average, one in every four days is rainy.
(b) Given that on a day she arrives late, what is the probability that it is a rainy day?

76. At an art exhibition there are 12 paintings of which 10 are original. A visitor selects a painting at random and before he decides to buy, he asks the opinion of an expert about the authenticity of the painting. The expert is right in 9 out of 10 cases on average.
(a) Given that the expert decides that the painting is authentic, what is the probability that this is really the case?
(b) If the expert decides that the painting is a copy, then the visitor returns

it and chooses another one; what is the probablity that his second choice is an original?

77. What is the conditional probability that a hand at poker (see Exercise 40) consists of spades, given that it consists of black cards?

Note: Of the 52 cards of an ordinary deck, the 13 spades and 13 clubs are black whereas the 13 hearts and 13 diamonds are red.

78. The probability p_k that a family has k children is given by $p_0 = p_1 = a$, $p_k = (1 - 2a)2^{-(k-1)}$ ($k \geq 2$). It is known that a family has two boys. What is the probability that:
(a) the family has only two children?
(b) the family has two girls as well?

79. An amplifier may burn one or both of its fuses if one or both of its tubes are defective. Let the events:
A_i: only tube i is defective ($i = 1, 2$);
A_3: both tubes are defective;
B_j: fuse j burns out ($j = 1, 2$);
B_3: both fuses burn out.
The conditional probabilities of burning the fuses (given the state of the tubes, i.e., the $P[B_j|A_i]$), appear in the following table:

i/j	B_1	B_2	B_3
A_1	0.7	0.2	0.1
A_2	0.3	0.6	0.1
A_3	0.2	0.2	0.6

Find the probabilities that:
(a) both tubes are defective given that both fuses were burnt out;
(b) only tube 2 is defective given that both fuses were burnt out;
(c) only tube 2 is defective given that at least one fuse was burnt out.
The events A_1, A_2, A_3 have probabilities 0.3, 0.2, 0.1, respectively.

CHAPTER 2

Distributions. Random Variables

Elements of Theory

1. A function (strictly measurable) $X(\omega)$ defined on the sample space $\Omega = \{\omega\}$ is called a *random* or *stochastic* variable.

A random variable is called *discrete* if it takes on (with positive probability) at most a countable set of values, that is, if there exists a sequence $x_1, x_2, \ldots$ with

$$P[X = x_i] = p_i > 0 \quad \text{and} \quad \sum_{i=1}^{\infty} p_i = 1. \tag{2.1}$$

The sequence $\{p_n, n = 1, \ldots\}$ defines the so-called probability (mass) function or frequency function of X.

A random variable X is called *continuous* (strictly, *absolutely continuous*) if for every real c there exists a function f (almost everywhere continuous) such that

$$P[X \leq c] = \int_{-\infty}^{c} f(x)\, dx. \tag{2.2}$$

The function f is called the *probability density function* or simply the *density* of X. It follows that:

2. The *distribution function* or *cumulative distribution function* of a random variable X is defined by

$$F(x) \equiv P[X \leq x] \quad \text{for every real } x.$$

Thus for a continuous random variable,

$$F(x) = \int_{-\infty}^{x} f(t)\, dt \quad \text{for every } x, \tag{2.3}$$

while for a discrete random variable

$$F(x) = \sum_{x_i \leq x} P[X = x_i].$$

From (2.3), it follows that if F is differentiable at x then

$$\frac{dF(x)}{dx} = f(x).$$

Moreover, the *probability differential* $f(x)\, dx$ can be interpreted as

$$f(x)\, dx = P[x < X < x + dx].$$

For a continuous random variable X, we have

$$P[\alpha < X < \beta] = P[\alpha \leq X \leq \beta] = P[\alpha < X \leq \beta] = P[\alpha \leq X < \beta],$$

i.e., no point carries positive probability and every set A of the real line of zero length carries zero probability.

3. The *main discrete distributions*
   (i) The *hypergeometric* (cf. (1.7)) with parameters N, n, and p;

$$P[X = r] = \binom{Np}{r}\binom{Nq}{n-r} \Big/ \binom{N}{n}, \qquad q = 1 - p. \tag{2.4}$$

   (ii) The *binomial* with parameters n (number of independent Bernoulli trials) and p (probability of success on each trial):

$$P[X = k] \equiv b(k; n, p) = \binom{n}{k} p^k q^{n-k}, \qquad k = 0, 1, \ldots, n, \quad q = 1 - p. \tag{2.5}$$

   (iii) The *Poisson distribution* with parameter λ:

$$P[X = k] \equiv p(k|\lambda) = e^{-\lambda}\frac{\lambda^k}{k!}, \qquad k = 0, 1, 2, \ldots. \tag{2.6}$$

   (iv) The *geometric* with parameter p (the number of Bernoulli trials required before or including the first success):

$$P[X = k] = pq^k, \qquad k = 0, 1, 2, \ldots,$$

or $\qquad P[X = k] = pq^{k-1}, \qquad k = 1, 2, 3, \ldots.$ $\tag{2.7}$

   (v) The *Pascal distribution* with parameters p and r (number of failures before the rth success in a sequence of Bernoulli trials):

$$P[X = k] = \binom{r + k - 1}{r - 1} p^r q^k = \binom{-r}{k} p^r (-q)^k, \qquad k = 0, 1, \ldots. \tag{2.8}$$

If r is simply positive (not necessarily an integer) then (2.8) is also called a *negative binomial* distribution.

4. The *main continuous distributions*

(i) *Uniform or rectangular* in the interval (α, β) with density

$$f(x) = \frac{1}{\beta - \alpha}, \qquad \alpha < x < \beta.$$

(ii) *Exponential* (or *negative exponential*) with density

$$f(x) = \theta e^{-\theta x}, \qquad x > 0.$$

(iii) *Normal* or *Gaussian* or *Gauss–Laplace* with parameters μ (mean) and σ^2 (variance) denoted by $N(\mu, \sigma^2)$ with density

$$f(x) = \frac{1}{\sigma\sqrt{2\pi}} \exp\left[-\frac{1}{2}\frac{(x - \mu)^2}{\sigma^2}\right], \qquad -\infty < x < \infty.$$

(iv) *Laplace distribution* (bilateral or double exponential) with density

$$f(x) = \tfrac{1}{2}e^{-|x|}, \qquad -\infty < x < \infty. \tag{2.9}$$

(v) β *distribution* with parameters $p > 0$ and $q > 0$ with density

$$\beta(x|p, q) = \frac{\Gamma(p + q)}{\Gamma(p)\Gamma(q)} x^{p-1}(1 - x)^{q-1}, \qquad 0 < x < 1,$$

$$= \frac{1}{B(p, q)} x^{p-1}(1 - x)^{q-1}. \tag{2.10}$$

where the Γ function is defined for every $p > 0$ by

$$\Gamma(p) = \int_0^\infty x^{p-1}e^{-x}\,dx,$$

and the B function is defined for every $p > 0$, $q > 0$ by

$$B(p, q) = \int_0^1 x^{p-1}(1 - x)^{q-1}\,dx.$$

(vi) The gamma (Γ) *distribution with* parameters $\lambda > 0$ (scale parameter) and $s > 0$ (shape parameter) with density

$$\gamma(x/\lambda, s) = \frac{\lambda^s}{\Gamma(s)} e^{-\lambda x} x^{s-1}, \qquad x > 0. \tag{2.11}$$

When $\lambda = 1/2$ and $2s = v =$ integer, the gamma (Γ) distribution is called a χ^2 (*chi-square*) *distribution with v degrees of freedom, denoted by* χ_v^2.

(vii) *Cauchy distribution* with density

$$f(x) = \frac{1}{\pi}\frac{1}{1 + x^2}, \qquad -\infty < x < \infty. \tag{2.12}$$

(viii) *Student's or t distribution* with v degrees of freedom. The distribution of the ratio $Z/\sqrt{\chi_v^2/v}$ where Z is $N(0, 1)$ independent of the χ_v^2. It has density

$$f_v(t) = \frac{\Gamma\left(\dfrac{v+1}{2}\right)}{\sqrt{\pi v}\,\Gamma\left(\dfrac{v}{2}\right)}\frac{1}{\left(1+\dfrac{t^2}{v}\right)^{(1+v)/2}}, \qquad -\infty < t < \infty.$$

(ix) *F or Snedecor's distribution* with m and n degrees of freedom. The distribution of the ratio $nx_m^2/m\chi_n^2$ where the χ_m^2 and χ_n^2 are independent. It has density

$$f(x) = \frac{\Gamma\left(\dfrac{m+n}{2}\right)}{\Gamma\left(\dfrac{m}{2}\right)\Gamma\left(\dfrac{n}{2}\right)}\left(\frac{m}{n}\right)^{m/2}\frac{x^{(m/2)-1}}{\left(1+\dfrac{m}{n}x\right)^{(m+n)/2}}, \qquad x > 0.$$

Exercises

1. Discrete Distributions

80. An urn contains 7 white balls numbered 1, 2, ..., 7 and 3 black balls numbered 8, 9, 10. Five balls are randomly selected, (a) with replacement, (b) without replacement.

For each of the cases (a) and (b) give the distribution:
   (I) of the number of white balls in the sample;
  (II) of the minimum number in the sample;
 (III) of the maximum number in the sample;
 (IV) of the minimum number of balls needed for selecting a white ball.

81. (*Continuation*). As in Problem 80 with the black balls numbered 1, 2, and 3.

82. A machine normally makes items of which 4% are defective. Every hour the producer draws a sample of size 10 for inspection. If the sample contains no defective items he does not stop the machine. What is the probability that the machine will not be stopped when it has started producing items of which 10% are defective.

83. One per thousand of a population is subject to certain kinds of accident each year. Given that an insurance company has insured 5,000 persons from the population, find the probability that at most 2 persons will incur this accident.

84. A certain airline company, having observed that 5% of the persons making reservations on a flight do not show up for the flight, sells 100 seats on a plane that has 95 seats. What is the probability that there will be a seat available for every person who shows up for the flight?

85. Workers in a factory incur accidents at the rate of two accidents per week. Calculate the probability that there will be at most two accidents, (i) during 1 week, (ii) during 2 weeks (iii) in each of 2 weeks.

86. Suppose that the suicide rate in a certain state is four suicides per one million inhabitants per month. Find the probability that in a certain town of population 500,000 there will be at most four suicides in a month. Would you find it surprising that during 1 year there were at least 2 months in which more than four suicides occurred?

87. A coin is tossed 30 times. For $n = 1, 2, \ldots, 20$, what is the conditional probability that exactly $10 + n$ heads appear given that the first 10 tosses resulted in heads. Show that the conditional probability that $10 + n$ tosses will result in heads, given that heads appeared at least ten times, equals

$$\frac{\binom{30}{10+n}\dfrac{1}{2^n}}{\sum_{k=0}^{20}\binom{30}{10+k}\dfrac{1}{2^k}}.$$

88. How many children should a family have so that with probability 0.95 it has at least a boy and at least a girl.

89. Show that the Poisson probabilities $p(k|\lambda)$ satisfy the recurrence relation

$$p(k|\lambda) = \frac{\lambda}{k}p(k-1|\lambda),$$

and hence determine the values of k for which the terms $p(k|\lambda)$ reach their maximum (for given λ).

2. Continuous Distributions

90. Verify that each of the following functions f is a probability density function and sketch its graph.

(a) $f(x) = 1 - |1 - x|$ for $0 < x < 2$

(b) $f(x) = \dfrac{1}{\pi}\dfrac{\beta}{\beta^2 + (x-a)^2}$ for $-\infty < x < \infty$,

(c) $f(x) = \dfrac{1}{2\sigma}e^{-(|x-\mu|)/\sigma}$ for $-\infty < x < \infty$,

(d) $f(x) = \tfrac{1}{4}xe^{-x/2}$ for $0 < x < \infty$.

91. The amount of bread (in hundreds of kilos) that a bakery sells in a day is a random variable with density

$$f(x) = \begin{cases} cx & \text{for } 0 \le x < 3, \\ c(6-x) & \text{for } 3 \le x < 6, \\ 0 & \text{otherwise.} \end{cases}$$

(i) Find the value of c which makes f a probability density function.

(ii) What is the probability that the number of kilos of bread that will be sold in a day is, (a) more than 300 kilos? (b) between 150 and 450 kilos?

(iii) Denote by A and B the events in (a) and (b), respectively. Are A and B independent events?

92. Suppose that the duration in minutes of long-distance telephone conversations follows an exponential density function;

$$f(x) = \tfrac{1}{5}e^{-x/5} \quad \text{for} \quad x > 0.$$

Find the probability that the duration of a conversation:
(a) will exceed 5 minutes;
(b) will be between 5 and 6 minutes;
(c) will be less than 3 minutes;
(d) will be less than 6 minutes given that it was greater than 3 minutes.

93. A number is randomly chosen from the interval $(0, 1)$. What is the probability that:
(a) its first decimal digit will be a 1;
(b) its second decimal digit will be a 5;
(c) the first decimal digit of its square root will be a 3?

94. The height of men is normally distributed with mean $\mu = 167$ cm and standard deviation $\sigma = 3$ cm.

(I) What is the percentage of the population of men that have height, (a) greater than 167 cm, (b) greater than 170 cm, (c) between 161 cm and 173 cm?

(II) In a random sample of four men what is the probability that:
(i) all will have height greater than 170 cm;
(ii) two will have height smaller than the mean (and two bigger than the mean)?

95. A machine produces bolts the length of which (in centimeters) obeys a normal probability law with mean 5 and standard deviation $\sigma = 0.2$. A bolt is called defective if its length falls outside the interval $(4.8, 5.2)$.
(a) What is the proportion of defective bolts that this machine produces?
(b) What is the probability that among ten bolts none will be defective?

96. Consider a shop at which customers arrive at random at a rate of twenty persons per hour. What is the probability that the time intervals between succesive arrivals will be:
(a) shorter than 3 minutes;
(b) longer than 4 minutes.
(c) Suppose that 10% of the customers buy a certain object. Find the distribution of the number of customers who buy an object in an hour.

97. If X is a continuous random variable with cumulative distribution function F and density function f, show that the random variable $Y = X^2$ is also continuous and express its cumulative distribution function and density in terms of F and f.

98. (*Continuation*). Find the density of $Y = X^2$ when X has:
(a) the normal distribution $N(\mu, \sigma^2)$;
(b) the Laplace distribution (see (2.9));
(c) the Cauchy distribution (see (2.12)).

99. As in Exercises 97 and 98 with $Y = |X|$.

100. The *lognormal distribution*. If the log X is normally distributed then X is said to have a lognormal distribution. Find its density.

CHAPTER 3

Expectation. Variance. Moments

Elements of Theory

1. The *expected value or expectation or mean value or simply the mean* of a random variable X, denoted by $E(X)$, is defined by

$$E(X) = \begin{cases} \sum_i x_i P[X = x_i] & \text{for a discrete } X, \\[2ex] \int_{-\infty}^{\infty} xf(x)\,dx & \text{for a continuous } X, \end{cases}$$

provided the series or the integral converge absolutely, in which case we say that $E(X)$ exists and write $E(X) < \infty$.

The *linearity property* of the expectation operation: If $E(X) < \infty, E(Y) < \infty$, then for any constants a and b we have

$$E[aX + bY] = aE(X) + bE(Y).$$

The mean value of a function of a random variable: Let

$$Y = g(X)$$

be a (measurable) function of X with frequency function f_X. Then

$$E(Y) = E[g(X)] = \begin{cases} \sum_i y_i f_Y(y_i), \\[2ex] \int_{-\infty}^{\infty} yf_Y(y)\,dy, \end{cases}$$

$$= \begin{cases} \sum_i g(x_i)f(x_i) & \text{for } X \text{ discrete,} \\[2mm] \int_{-\infty}^{\infty} g(x)f(x)\, dx & \text{for } X \text{ continuous,} \end{cases} \qquad (3.1)$$

where f_Y denotes the frequency function of Y; (3.1) permits the computation of $E(Y)$ by means of f_X *without finding first* f_Y.

2. *Moments.* The moment of order r about the origin of a random variable X is defined by

$$\mu_r' \equiv E(X^r), \qquad r = 1, 2, \dots .$$

The first moment μ_1' coincides with the mean $E(X) \equiv \mu$.

The *central moment* of order r is defined by

$$\mu_r \equiv E(X - \mu)^r \qquad r = 2, 3, \dots .$$

The *variance* of X is the second (order) central moment, i.e.,

$$E(X - \mu)^2 = E(X^2) - \mu^2. \qquad (3.2)$$

We write V or Var for variance. The positive square root of $\mathrm{Var}(X)$ is called the *standard deviation* of X, usually denoted by σ.

The moments of an integer-valued positive random variable, such as the binomial, the Poisson, etc., are more conveniently computed by means of the so-called *factorial moments*.

The *factorial moment* π_r of the random variable X is defined by

$$\pi_r = E[X(X - 1)\dots(X - r + 1)] = E[(X)_r].$$

Thus

$$\mathrm{Var}(X) = \pi_2 + \mu - \mu^2. \qquad (3.3)$$

3. *Quantiles and percentiles.* For every p $(0 < p < 1)$, the p quantile point or the $100p$ percentile point, X_p say, of a random variable X (or its distribution) with distribution function F, is defined as a solution of

$$F(X_p-) \le p \le F(X_p).$$

For a continuous and increasing F, X_p is uniquely defined by

$$F(X_p) = p.$$

For $p = 0.5$ we have the *median* $X_{0.5}$. $X_{0.25}$ is called the *first quartile* and $X_{0.75}$ the *third quartile*. The difference $X_{0.75} - X_{0.25}$ is called *interquartile range* and can be used as a measure of scatter of the distribution.

4. *Probability or factorial moment generating function.* This is useful mainly for integer-valued positive random variables and is defined by

$$P(t) \equiv E(t^X) = \sum_{k=0}^{\infty} t^k P[X = k] = \sum_{k=0}^{\infty} p_k t^k;$$

it exists at least for $|t| \le 1$.

The rth factorial moment π_r is given by

$$\pi_r = \left[\frac{d^r P(t)}{dt^r}\right]_{t=1} = P_{(1)}^{(r)},$$

so that, by (3.3),

$$E(X) = P'(1), \qquad \text{Var}(X) = P''(1) + P'(1) - [P'(1)]^2.$$

The probabilities $p_k = P[X = k]$ are given by

$$p_k = \frac{1}{k!}\left[\frac{d^k P(t)}{dt^k}\right]_{t=0}.$$

5. *Moment generating function* of a random variable X defined by

$$M(t) \equiv E(e^{tX}) = \begin{cases} \sum_i e^{tx_i} P[X = x_i] & \text{(discrete } X\text{)}, \\ \int_{-\infty}^{\infty} e^{tx} f(x)\, dx & \text{(continuous } X\text{)}. \end{cases}$$

If $M(t)$ exists, i.e., there exists a $\delta > 0$ such that $E(e^{tX}) < \infty$ for all $|t| < \delta$, then all the moments $E(X^r)$ $(r = 1, 2, \ldots)$ exist and can be obtained from

$$E(X^r) = \left[\frac{d^r M(t)}{dt^r}\right]_{t=0} = M^{(r)}(0).$$

Exercises

1. Theoretical Exercises

101.* Show that the integral $\int_{-\infty}^{+\infty} |x - m| f(x)\, dx$ becomes minimum when m is the median of the distribution with density f.

102. Let $f(x)$ denote the density function of the random variable X. Suppose that X has a symmetric distribution about a, that is, $f(x + a) = f(a - x)$ for every x. Show that the mean $E(X)$ equals a, provided it exists.

103.* *Cauchy–Schwarz inequality.* Show that $E^2(XY) \le E(X^2)E(Y^2)$ provided the second moments exist.

104. Show that $|E(X)| \le E(|X|)$.

105. For every set A of real numbers, we define the indicator function by

$$I_A(x) = \begin{cases} 1 & \text{if } x \in A, \\ 0 & \text{if } x \notin A. \end{cases}$$

Show that

$$P(A) = E[I_A(X)] = \int_A dF_X(x).$$

106. If $E(X) = E(X^2) = 0$, show that $P(X = 0) = 1$.
Hint: Use Chebyshev's inequality. See (8.5).

107. Show that the mean μ of a random variable X has the property

$$\min_c E(X - c)^2 = E(X - \mu)^2 = V(X).$$

108.* *The mean value geometrically.* Show that for a continuous random variable X with density function f and cumulative distribution function F

$$\mu = E(X) = \int_0^{+\infty} [1 - F(x)]\, dx - \int_{-\infty}^0 F(x)\, dx.$$

Consequently, in the graph below $\mu = \text{area}(A) - \text{area}(B)$.

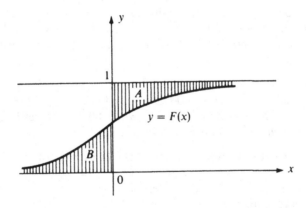

109. If the nth moment of the random variable X with distribution function F exists, show that

$$E(X - c)^k = k \int_c^\infty (x - c)^{k-1}[1 - F(x)]\, dx$$

$$- k \int_{-\infty}^c (x - c)^{k-1} F(x)\, dx, \qquad 1 \le k \le n.$$

110. Show that the first n moments determine the first n central moments and, conversely, that the first n central moments with the mean determine the first n moments.

111. If X is bounded, i.e., there is a constant $k < \infty$ such that $P[|X| \le k] = 1$, then X has moments of every order.

112.* If the n-order moment μ'_n exists, then show that there exist all μ'_k $(k = 1, 2, \ldots, n - 1)$.

113. Show that a necessary condition for the mean of a random variable with distribution function F to exist is that

$$\lim_{x \to -\infty} xF(x) = \lim_{x \to +\infty} x[1 - F(x)] = 0.$$

2. Mean and Variance

114.* In a lottery, n numbers are selected from the N numbers $1, 2, \ldots, N$. Find the variance of the sum S_n of the selected numbers.

115. An urn contains N_1 white balls and N_2 black balls; n balls are drawn at random, (a) with replacement, (b) without replacement. What is the expected number of white balls in the sample?

116. A student takes a multiple-choice test consisting of two problems. The first one has 3 possible answers and the second one has 5. The student chooses, at random, one answer as the right one from each of the two problems. Find:
(a) the expected number, $E(X)$, of the right answers X of the student;
(b) the Var(X).
Generalize.

117. In a lottery that sells 3,000 tickets the first lot wins $1,000, the second $500, and five other lots that come next win $100 each. What is the expected gain of a man who pays 1 dollar to buy a ticket?

118. A die is thrown until the result "ace or even number" appears three times. Find the expected number of throws:
(a) in one performance of the experiment;
(b) in ten repetitions.

119. A pays 1 dollar for each participation in the following game: three dice are thrown; if one ace appears he gets 1 dollar, if two aces appear he gets 2 dollars and if three aces appear he gets 8 dollars; otherwise he gets nothing. Is the game fair, i.e., is the expected gain of the player zero? If not, how much should the player receive when three aces appear to make the game fair?

120. A player has 15 dollars. If heads appear on the first toss he receives 1 dollar and he withdraws from the game. If tails appear he bets 2 dollars and if he wins the second toss he withdraws; otherwise he continues playing on a bet of 4 dollars. If he loses, he bets the remaining amount of 8 dollars. What is the expected gain of the player?

121. Of the parcels mailed abroad 10% never reach their destination. Two books may be sent separately or in a single parcel. Each book is worth $2. The postage for each book sent separately is 10¢ and for both books in a single

parcel 15¢. For each of the two ways of mailing, find:
  (a) the probability that both books reach their destination;
  (b) the probability that at least one book reaches its destination;
  (c) the expected net (after postage) value of the commodities reaching their destinatination.
For each of the three criteria (a), (b), and (c), which way of mailing is preferable?

122. (a) Two indentical coins are thrown once in such a way that if one shows heads so does the other (dependent). Let $P[\text{heads}] = p$ and X be the total number of heads that appear. Find the mean and the variance of the random variable X.

(b) Two different coins with probabilities of heads p_1 and p_2 are thrown independently. Find the mean and the variance of the random variable X as defined in (a).

123. *Mixed distribution.* Let X be a random variable with distribution function

$$F(x) = \begin{cases} 1 - 0.8e^{-x} & \text{for } x \geq 0, \\ 0 & \text{for } x < 0. \end{cases}$$

Plot the graph of $F(x)$ and then evaluate $E(X)$. Give an example of a random variable X that has the above distribution.

124. If the random variable X is $N(\mu, 1)$, show that the random variable $Y = [1 - \Phi(X)]/\varphi(X)$, where Φ and φ denote the distribution function and the density of $N(0, 1)$, respectively, has mean value $1/\mu$.

125. The truncated Poisson distribution with the zero class missing has probability function

$$P(X = k) = \frac{\lambda^k}{(e^\lambda - 1)k!}, \qquad k = 1, 2, \ldots.$$

Find $E(X)$ and $\text{Var}(X)$.

CHAPTER 4

General Problems

126. Three players A, B, and C play a game as follows: At the first stage A and B play against each other while C stays out of the game. The winner plays against C at the second stage. Then the winner of the second game plays against the last loser, and so on. The winner is declared the one who wins two games in succession. Find the elements (simple events) of the sample space Ω of all possible outcomes of the game. If the probability of each player winning a game is 1/2, determine the probabilities of the corresponding possible outcomes and show that their sum equals one.

127. A symmetrical coin is tossed until the same result appears twice in succession. Describe the sample space in terms of the results H ("heads") and T ("tails"). Find the probabilities of the events:

(a) the experiment terminates earlier or at the seventh trial;

(b) an even number of trials is required.

What is the expected number of trials?

128. A pair of dice is tossed six times. What is the probability that all the faces will appear twice?

129. A deck of 52 cards is divided among four players so that each of them gets 13 cards (game of bridge). Find the probability that at least one player will have a complete suit (spades, hearts, diamonds, clubs).

130. Two coins C_1 and C_2 have a probability of falling heads p_1 and p_2, respectively. You win a bet if in three tosses you get at least two heads in succession. You toss the coins alternately starting with either coin. If $p_1 > p_2$, what coin would you select to start the game?

131. Ten pairs of shoes are in a closet. Four shoes are selected at random. Find the probability that there will be at least one pair among the four shoes selected.

132. Let N cells be numbered 1, 2, ..., N. We randomly throw balls into them. The process is continued until a ball falls in the cell bearing the number 1. What is the probability that:

(a) n throws will be necessary?

(b) more than n throws will be necessary?

133. (*Continuation*). The process terminates when any one of the cells receives two balls.

(a) What is the probability that more than n throws will be necessary?

(b) What is the expected number of throws?

134. From a usual deck of cards we draw one card after another. What is the probability that:

(a) the nth card will be the first ace?

(b) the first ace will appear among the first n cards?

135. The 52 cards of an ordinary deck of cards are placed successively one after the other and from left to right. Find the probability that the thirteenth spade will appear before the thirteenth diamond.

136. Three balls are drawn at random one after another and without replacement from n balls numbered 1, 2, ..., n. Find the probability that the first will bear a number smaller than that of the second ball.

137. An urn contains m kinds of objects all in the same proportion. Objects are drawn with replacement, one after another, until each kind appears at least once. What is the probability p_v that v objects will be required?

138. A newspaper dealer gets n papers every day for sale. The number X of papers sold is a random variable following the Poisson distribution with parameter λ. For each paper sold he earns 1 cent, for each unsold paper he loses b cents. Let Y denote the net gain of the dealer. Find:

(a) $E(Y)$;

(b) How many newspapers should he get every day in order to maximize his profit?

139. Three numbers are selected at random one after another and without replacement from n numbers 1, 2, ..., n. What is the probability that the first number drawn will be the smallest and the second number the largest?

140. The events A, B, and C are independent with $P(A) = 0.2$, $P(B) = 0.3$, and $P(C) = 0.1$. Find the probability that at least two will occur among the three events.

141. We throw four coins simultaneously and we repeat it once more. What is the probability that the second throw will result in the same configuration as the first throw when the coins are, (a) distinguishable, (b) indistinguishable.

142. An urn contains w white balls and b black balls. We draw balls successively and without replacement one after another until a white ball appears for the first time. Let X be the required number of draws. Find:

(a) the distribution of X;

(b) $E(X)$.

(c) Using the result of (a) show the identity

$$1 + \frac{n-w}{n-1} + \frac{(n-w)(n-w-1)}{(n-1)(n-2)} + \cdots + \frac{(n-w)\ldots 2.1}{(n-1)\ldots(w+1)w} = \frac{n}{w},$$

where $n = w + b$.

143. (*Continuation*). Let $w = b = N$ and draw balls successively and without replacement one after another until all the $2N$ balls are drawn. What is the probability that in a stage of the experiment the same number of white and black balls has been drawn?

144.* Let $B_1, \ldots, B_v, \ldots$, be a partition of the sample space Ω and $P(C) > 0$. Show the following theorem of total probability:

$$P[A|C] = \sum_{j=1}^{\infty} P[B_j|C]P[A|B_jC].$$

145. Each of the three Misses T. C—Rena, Nike, and Galatea—wants to accompany me on a trip. Since I cannot take all of them, I play the following game: I say to them, "I will think of one of the numbers 1, 2, 3. Rena, you guess first. If you guess right you will come with me; otherwise Nike will guess next. If she guesses right she will come with me. Otherwise Galatea will accompany me". Galatea complains that the game is unfair. Is she right? Suppose every one of them writes down on a piece of paper the number which she guesses I thought of and I follow the above sequence checking Rena's number first, etc. This is repeated until one is chosen to come with me. Galatea and Nike protest this time. Are they justified?

146. A die is thrown v times. Find the probability that each of k ($1 \leq k \leq 6$) given faces of the die appears at least once.

147. The folded normal distribution has density

$$f(x) = c\frac{1}{\sqrt{2\pi}}e^{-x^2/2}, \qquad x > 0.$$

Determine the constant c and then evaluate the mean of the distribution.

148. Suppose the lifetime of an electric lamp of a certain type obeys a normal law with mean $\mu = 180$ hours and standard deviation $\sigma = 20$. In a random sample of four lamps:

(a) What is the probability that all four lamps have a lifetime greater than 200 hours?

(b) The random sample of four lamps is placed in an urn and then we randomly draw one lamp from it; what is the probability that the lamp will have a lifetime greater than 200 hours?

Generalize the conclusion.

149.* The *game of "craps"* is played as follows. The gambler throws two dice. If at the first throw he gets 7 or 11 he wins, and if he gets 2, 3, or 12 he loses. For each of the other sums the game is continued in two ways:

(a) the gambler continues throwing the two dice until he wins with a 7 or he loses with the results of the outcome of the first throw;

(b) the gambler continues until he loses with 7 or wins with the result of the first throw.

What is the probability of the gambler winning in cases (a) and (b)?

150. Two gamblers A and B agree to play as follows. They throw two dice and if the sum S of the outcomes is < 10 B receives S dollars from A, otherwise B pays A x dollars. Determine x so that the game is fair.

151. An urn contains n white balls and n red balls. We draw two balls at first, and then another two, and so on, until all the balls are drawn. Find the probability that each of the selected pairs consists of a white ball and a red ball.

152. Each of two urns A and B contains n balls numbered 1 to n. We draw one ball from each of the urns. Find the probability that the ball drawn from A bears a number smaller than that drawn from B.

153. *The rumor mongers* (see Feller, 1957, p. 55). In a town of $N + 1$ inhabitants, a person tells a rumor to a second person, who in turn repeats it to a third person, and so on. At each step the recipient of the rumor is chosen at random from the N inhabitants available.

(i) Find the probability that the rumor will be told n times without, (a) returning to the originator, (b) being repeated to any person.
Do the same problem when at each step the rumor is told to k persons.

(ii) In a large town where the rumor mongers constitute $100p\%$ of the population, what is the probability that the rumor does not return to the originator?

154.* *Pascal's problem.* In the game of tossing a fair coin, the first one to obtain n successes (heads or tails) wins. Show that the game is fair (i.e., each gambler has a probability of winning equal to $1/2$). Suppose that the game was interrupted when the first gambler had won k tosses and the second gambler had won m tosses ($0 \leq k, m < n$). Calculate for each gambler the probability of winning if the game is continued. Hence deduce how the stake should be divided after interrupting the game.

155.* *Chebyshev's problem.* Find the probability that a given fraction m/n (where m and n are integers) is irreducible.

156.* Let $\varphi(n)$ denote Euler's function, i.e., the number of (positive) integers which are primes relative to n and smaller than n. Using a probabilistic argument, show that

$$\varphi(n) = n \prod_{p/n} \left(1 - \frac{1}{p}\right),$$

where the product extends over all prime divisors p of n.

157.* Assume that the number of insect colonies in a certain area follows the Poisson distribution with parameter λ, and that the number of insects in a colony has a logarithmic distribution with parameter p. Show that the total number of insects in the area is a negative binomial with parameters $q = 1 - p$ and $-\lambda/\log(1 - p)$.

158. Electronic tubes come in packages, each containing N tubes. Let p_k denote the probability that a package contains k defective tubes ($0 \le k \le m$). A sample of n tubes is taken from a package and it is observed that $r \le m$ tubes are defective.

(a) What is the probability that the selected package actually contains k ($\ge r$) defective tubes?

(b) If according to customer demand a package is considered not acceptable whenever it contains $d \ge r$ defective tubes, what is the probability that the package is not accepted?

159. The probability p_n that n customers visit a supermarket in one day is $p_n = p^n q$, $n = 0, 1, \ldots$. Two out of three customers, on average, buy a certain type of item. The probability that an item is defective is $1/4$.

(a) What is the probability that a customer buys a nondefective item?

(b) Given that k nondefective items were sold, show that the conditional probability a_n that n customers visited the shop is given by

$$a_n = \binom{n}{k} p^{n-k}(2 - p)^{k+1}/2^{n+1}.$$

160. A manufacturer sells an item for \$1. If the weight of the item is less than W_0, it cannot be sold and it represents a complete loss. The weight W of an item follows the normal distribution $N(\mu, 1)$; the cost c per item is given by $c = \alpha + \beta W$ (α, β positive constants). Determine the mean μ so that the expected profit is maximized.

161. A clerk lives at A and works at C and he starts work at 9 a.m. The clerk always takes the train from A to B which is supposed to reach B at 8:40 a.m. Buses from B leave for C every 15 minutes and the bus which leaves at 8:45 a.m. is supposed to arrive at 8:56 a.m.

The train on average experiences delays of 2 minutes and has a standard deviation of 4 minutes. The bus always leaves on time, but arrives on average after a 2-minute delay and a standard deviation of 3 minutes. What is the probability that the cleark arrives late? The clerk's employer drives to his office; he leaves at 8:45 a.m. and the driving time to the office has a mean value of 12 minutes and a standard deviation of 2 minutes. Find the probability that:

(a) both, clerk and employer, arrive late;

(b) the employer arrives earlier than the clerk.

Assume that the distributions involved are normal.

162.* *Quality control.* In a continuous manufacturing process the percentage of defective articles is p. To maintain the quality of the product at a certain standard, the articles are examined, one by one, until a sequence of fixed length r appears with no defective article. Then total inspection terminates and only a certain fraction f of the product is chosen at random for inspection until a defective article appears when the above process of 100% inspection continues. Under the described sampling scheme find:

(a) the probability of a defective sequence, i.e., that a defective article is not followed by r successive good articles;

(b) the expected number of articles in a defective sequence. Hence deduce the expected number, (i) of defective sequences, (ii) of inspected articles, after a defective article;

(c) the expected proportion of the product subject to inspection;

(d) the expected percentage ρ of defective articles going out for sale provided that any discovered defective article is replaced by a good one;

(e) for given values of f and r, what value p^*, say, of p maximizes ρ.

163.* *The dilemma of the convict.* Three convicts A, B, and C appeal for parole and the board decides to free two of them. The convicts are informed of this decision but are not told who the two to be set free are. The guard, who is A's friend, knows which convicts are going to be set free. A sees that it would not be right to ask the guard about his own fate, but thinks that he might ask for the name of one of the other two who will be set free. He supposes that before he asks the probability of his being set free is 2/3, and when the guard answers that B is pardoned, for example, the probability that he will be set free is lessened to 1/2, because either A and B will be set free or B and C. Having thought about this, A is afraid to ask the guard. Is his fear justified?

164.* *Collection of coupons.* In each box of a given product there is a coupon with a number from 1 to 6. If a housewife succeeds in getting the series 1–6 she receives a free box of the product. How many boxes must she buy, on average, before she gets a free one?

165.* In a row of 19 seats in an amphitheatre 10 male students and 9 female students sit at random. Find the expected number of successive pairs of seats in which a male student and a female student sit.

166.* A, B, and C are about to play to following fatal game. Each one has a gun and will shoot at his target—A, B, and C taking turns at each other in that order—and it is assumed that any one who is hit will not be shot at again. The shots, one after another, continue until one of the players remains unhit. What strategy should A follow?

167. Samuel Pepys (who was about to place a bet) asked Newton which of the following events A, B, or C is more probable: A—at least one six when 6 dice are thrown; B—at least two sixes when 12 dice are thrown; and C—at least three sixes when 18 dice are thrown. Which answer (the correct one) did Newton give?

168.* *Birthday holidays.* The Worker's Legal Code in Erehwon specifies as a holiday any day during which at least one worker in a certain factory has a birthday. All other days are working days. How many workers must the factory employ so that the number of working man-days is maximized during the year?

169. *Bertrand's paradox.* A chord AB is randomly chosen in a circle of radius r. What is the probability that the length of AB is less than r?

170.* *The neophyte at the horseraces.* At a horserace a neophyte better, who is about to bet a certain amount, wants to choose the best horse. It is assumed that no two horses are the same. The better looks at the horses as they pass by, one after another, and he can choose as the winner any one of the horses, but he cannot bet on a horse he has already let pass. Moreover, he can tell whether any horse passing by is better or worse compared with the preceding ones. Suppose n horses take part in the race. How should he proceed to choose the horse to bet on in order to maximize his probability of winning? What proportion of horses should he wait to pass by before he makes his bet when n is large? (It is assumed that horses parade in front of the neophyte in a random order.)

PART II

ADVANCED TOPICS

CHAPTER 5

Multivariate Distributions

Elements of Theory

1. *Multidimensional or vector random variable*: A real-valued vector function defined on the sample space Ω (more strictly, measurable with respect to the σ algebra on Ω).

The following definitions concerning bivariate distributions extend easily to more than two variables.

2. The *joint distribution function* of the random variables X and Y is defined by

$$F(x, y) \equiv P[X \leq x, Y \leq y] \qquad \text{for every point} \quad (x, y) \in R^2.$$

3. The distribution of the random pair (X, Y) is called *discrete* if there exists a denumerable set of points (x_i, y_j) such that

$$P[X = x_i, Y = y_j] = p_{ij} > 0,$$

and

$$\sum_i \sum_j p_{ij} = 1.$$

The probabilities p_{ij} $(i = 1, 2, \ldots, j = 1, 2, \ldots)$ define the so-called joint probability distribution or frequency function of X and Y. Clearly we have

$$F(x, y) = \sum_{\substack{x_i \leq x \\ y_j \leq y}} p_{ij}.$$

4. The distribution of the pair (X, Y) is called continuous if there exists a function $f(x, y)$ such that for every (x, y) the joint distribution function $F(x, y)$

of X and Y can be written as

$$F(x, y) = \int_{-\infty}^{x} \int_{-\infty}^{y} f(u, v) \, du \, dv.$$

The function $f(x, y)$ is called the *joint density function* of X and Y. In this case, for every continuity point (x, y) of f (almost all points are such), we have

$$f(x, y) = \frac{\partial^2 F(x, y)}{\partial x \, \partial y} = \lim_{\substack{\Delta x \to 0 \\ \Delta y \to 0}} \frac{P[x < X \le x + \Delta x, \, y < Y \le y + \Delta y]}{\Delta x \, \Delta y}.$$

5. *Marginal distributions.* The distribution of X, as obtained from the joint distribution of X and Y, is referred to as the *marginal* distribution of X. Similarly for Y. Thus for jointly discrete variables the marginal probability distribution of X is given by

$$p_i = P[X = x_i] = \sum_j p_{ij}, \tag{5.1}$$

while for continuous variables the marginal density f_X of X is given by

$$f_X(x) = \int_{-\infty}^{\infty} f(x, y) \, dy. \tag{5.2}$$

6. *Conditional distributions.* The conditional distribution function of X, given $Y \in B$ with $P[Y \in B] > 0$, is defined by

$$F_X(x|Y \in B) = P[X \le x | Y \in B] = \frac{P[X \le x, Y \in B]}{P[Y \in B]}. \tag{5.3}$$

The *conditional probability distribution* of X, given $Y = y_j$, is defined by

$$f_X(x_i|Y = y_j) = \frac{P[X = x_i, Y = y_j]}{P[Y = y_j]} = \frac{p_{ij}}{q_j}, \tag{5.4}$$

where we set $q_j = P[Y = y_j]$ $(j = 1, 2, \dots)$.

The *conditional density* of X, given $Y = y$, for the continuous pair (X, Y) is defined by

$$f_X(x|Y = y) = \frac{f(x, y)}{f_Y(y)} = \lim_{\Delta y \to 0} f_X(x|y < Y \le y + \Delta y)$$

$$= \lim_{\substack{\Delta x \to 0 \\ \Delta y \to 0}} \frac{P[x < X \le x + \Delta x, \, y < Y \le y + \Delta y]}{\Delta x \cdot P[y < Y \le y + \Delta y]}. \tag{5.5}$$

Hence, and by virtue of (5.1) and (5.2), we deduce the following.

7. *Formula of total probability for random variables:*

$$p_i = P[X = x_i] = \sum_j P[X = x_i | Y = y_j] P[Y = y_j] = \sum_j f_X(x_i | Y = y_j) q_j,$$

$$f_X(x) = \int_{-\infty}^{\infty} f_X(x|Y = y) f_Y(y) \, dy.$$

Similarly, we obtain the

8. *Generalized Bayes formula:*

$$f(x_k | Y = y_j) = \frac{p_k f_Y(y_j | X = x_k)}{\sum_i f_Y(y_j | X = x_i) p_i};$$

(5.6)

$$f_X(x | Y = y) = \frac{f_X(x) f_Y(y | X = x)}{\displaystyle\int_{-\infty}^{\infty} f_Y(y | X = x) f_X(x)\, dx}.$$

9. The expected or mean value of a function g of two random variables is defined by

$$E\{g(X, Y)\} = \begin{cases} \displaystyle\sum_i \sum_j g(x_i, y_j) p_{ij} & \text{(for discrete)}, \\[2mm] \displaystyle\int_{-\infty}^{\infty} \int_{-\infty}^{\infty} g(x, y) f(x, y)\, dx\, dy & \text{(for continuous)}. \end{cases}$$

We say it exists if the (double) series or integral is absolutely convergent.

10. The mixed central moment of orders $r + s$ of X and Y is defined by

$$\mu_{xs} = E[(X - \mu_x)^x (Y - \mu_y)^s],$$

where $\mu_x = E(X)$, $\mu_y = E(Y)$, and e.g., for continuous (X, Y), $E(X)$, is given by

$$E(X) = \int_{-\infty}^{\infty} \int_{-\infty}^{\infty} xf(x, y)\, dx\, dy = \int_{-\infty}^{\infty} x \left(\int_{-\infty}^{+\infty} f(x, y)\, dy \right) dx$$

$$= \int_{-\infty}^{+\infty} xf_x(x)\, dx.$$

11. The mixed central moment of the second order

$$\mu_{11} = E(X - \mu_x)(Y - \mu_y) \equiv \text{Cov}(X, Y)$$

is called the *covariance* of X and Y.

The number ρ $(-1 \le \rho \le 1)$, defined by

$$\rho(X, Y) = \frac{\text{Cov}(X, Y)}{\sigma_x \sigma_y},$$

where $\sigma_x^2 = \text{Var}(X)$, $\sigma_y^2 = \text{Var}(Y)$, is called the *correlation coefficient* between X and Y.

The following properties can easily be established. For all constants a, b, c, and d we have:

(i) $\text{Cov}(X + a, Y + b) = \text{Cov}(X, Y)$ (invariance under translations);

(ii) $\text{Cov}(cX, dY) = cd\, \text{Cov}(X, Y)$;

(iii) $\rho(aX + B, cY + d) = \text{sg}(ac)\rho(X, Y)$ where

$$\text{sg}(x) = \begin{cases} 1 & \text{for } x > 0, \\ -1 & \text{for } x < 0 \end{cases}$$

(iv) $\text{Var}(X \pm Y) = \text{Var}(X) + \text{Var}(Y) \pm 2\,\text{Cov}(X, Y)$.

12. The *regression* (*function*) of Y on X, $m_2(x)$ say, is defined as the conditional expected value of Y given $X = x$, that is,

$$m_2(x) = E(Y|X = x) = \int_{-\infty}^{\infty} y f_Y(y|X = x)\, dy = \frac{\int_{-\infty}^{\infty} y f(x, y)\, dy}{f_X(x)}.$$

The curve $y = m_2(x)$ is called the mean regression curve of Y on X. Similarly, we can define the regression of X on Y.

13. The *dispersion or variance–covariance* matrix of a random vector $X = (X_1, \ldots, X_n)$, usually denoted by Σ, is defined by

$$\Sigma = (\sigma_{ij}) \quad \text{where} \quad \sigma_{ij} = \text{Cov}(X_i, X_j), \quad i, j = 1, \ldots, n. \tag{5.7}$$

We shall write $D(X) = \Sigma$. If $a = (a_1, \ldots, a_n)'$ is a vector of constants then

$$\text{Var}(a_1 X_1 + \cdots + a_n X_n) = \text{Var}(a'X) = a'D(X)a = a'\Sigma a = \sum_{i=1}^{n} \sum_{j=1}^{n} a_i a_j \sigma_{ij}.$$

If the rank r of Σ is less than n, then the distribution of X is called singular. This is equivalent to the distribution being concentrated in an r-dimensional subspace of the n-dimensional Euclidean space E^n. For example, if $n = 2$ and $r = 1$ the distribution is concentrated on a straight line in the (X_1, X_2) plane and $|\rho(X_1, X_2)| = 1$.

14. The random variables $X_1, \ldots, X_n$ are called *completely stochastically independent* or simply *independent* if one of the following holds:
(I) For all Borel sets of the real line $A_1, \ldots, A_n$

$$P[X_1 \in A_1, \ldots, X_n \in A_n] = P[X_1 \in A_1] \ldots P[X_n \in A_n].$$

(II) The joint distribution function F of $X_1, \ldots, X_n$ can be written as

$$F(x_1, \ldots, x_n) = F_1(x_1) \ldots F_n(x_n) \quad \text{for every } (x_1, \ldots, x_n),$$

where F_i denotes the marginal distribution function of X_i.
(III) The joint frequency (density) function f can be written as

$$f(x_1, \ldots, x_n) = f_1(x_1) \ldots f_n(x_n),$$

where f_i denotes the marginal frequency function of X_i.

15. *The main multivariate distributions*
(a) *Discrete distributions*

(i) *Double hypergeometric:* R, W, B integers $N = R + B + W$. Then

$$p_{ij} = \binom{W}{i}\binom{R}{j}\binom{B}{n-i-j}\bigg/\binom{N}{n}, \qquad 0 \le i+j \le n,$$

(a generalization of the simple hypergeometric, see Exercise 175).

(ii) *Multinomial (k dimensional)*

$$P[X_1 = n_1, \ldots, X_k = n_k] = \frac{n!}{n_1! \ldots n_k! \, n_{k+1}!} p_1^{n_1} \cdots p_k^{n_k} p_{k+1}^{n_{k+1}}, \qquad (5.8)$$

where it was set

$$n_{k+1} = n - (n_1 + \cdots + n_k), \qquad p_{k+1} = 1 - (p_1 + \cdots + p_k).$$

(iii) *Negative multinomial (k dimensional)*

$$P[X_1 = n_1, \ldots, X_k = n_k] = \left(1 + \sum_{j=1}^{k+1} \theta_j\right)^{-s-v} \frac{\Gamma(s+v)}{\Gamma(s)} \prod_{j=1}^{k+1} \frac{\theta_j^{n_j}}{n_j!}, \qquad (5.9)$$

where $s > 0$, $\theta_j > 0$ $(n_j = 0, 1, 2, \ldots)$, and $n_{k+1} = n - (n_1 + \cdots + n_k)$.

(b) *Continuous distributions*

(i) Uniform in a bounded set S of E^n with density

$$f(x) = c, \qquad x \in S,$$

where c^{-1} = measure (length, area, volume, etc.) of S.

(ii) *n*-dimensional (nonsingular) normal $N(\mu, \Sigma)$ with density

$$f(x) = [(2\pi)^p |\Sigma|]^{-1/2} \exp[-\tfrac{1}{2}(x - \mu)' \Sigma^{-1}(x - \mu)], \qquad (5.10)$$

where $\mu = (\mu_1, \ldots, \mu_n)' = E(X)$ denotes the mean vector and Σ denotes the (positive definite) covariance matrix of the normal random vector X with density (5.10).

(iii) *Dirichlet* distribution with density

$$f(x_1, \ldots, x_k) = \frac{\Gamma(n_1 + \cdots + n_{k+1})}{\Gamma(n_1) + \cdots + \Gamma(n_{k+1})} \prod_{i=1}^{k+1} x_i^{n_i - 1}, \qquad \sum_{i=1}^{k+1} x_i = 1, x_i \ge 0. \qquad (5.11)$$

Exercises

171. The joint distribution of (X, Y) is defined by $P[X = 0, Y = 0] = P[X = 0, Y = 1] = P[X = 1, Y = 1] = 1/3$.

(a) Find the marginal distribution functions of X, Y.

(b) Examine whether the points $P_1: (-1/2, 0)$, $P_2: (0, 1)$ are continuity points of $F(x, y)$.

172. Show that the function

$$F(x, y) = \begin{cases} 0 & \text{for } x + y < 1, \\ 1 & \text{for } x + y \ge 1, \end{cases}$$

is not a joint distribution function.

173. We consider a family with two children. Let $X_n = 1$ if the nth child is a boy for $n = 1, 2$ and $X_3 = 1$ if there is only one boy, otherwise $X_i = 0$ ($i = 1, 2, 3$). Show that the X_i are pairwise independent but not completely independent.

174. X and Y have the joint density

$$f(x, y) = cx^{n_1-1}(y - x)^{n_2-1}e^{-y} \qquad \text{for} \quad 0 < x < y < \infty.$$

Find (a) the constant c, (b) the marginal distributions of X and Y.

175. The k-dimensional hypergeometric distribution has probability function

$$P[X_1 = n_1, \ldots, X_k = n_k] = \binom{Np_1}{n_1} \cdots \binom{Np_{k+1}}{n_{k+1}} \Big/ \binom{N}{n},$$

where

$$n_{k+1} = n - (n_1 + n_2 + \cdots + n_k),$$

$$p_{k+1} = 1 - (p_1 + p_2 + \cdots + p_k), \qquad n_i > 0, \quad p_i > 0.$$

Show that the conditional distribution of X_k given $X_1, \ldots, X_{k-1}$ is hypergeometric.

176. If $X_1, X_2, \ldots, X_k$ are distributed according to the multinomial distribution, the conditional distribution of X_1, given $X_2 = n_2, \ldots, X_k = n_k$, is binomial with parameters $n - (n_2 + \cdots + n_k)$ and $p_1/(p_1 + p_{k+1})$.

177. If $X_1, X_2, \ldots, X_k$ have a Dirichlet distribution, show that the conditional distribution of X_k/S_k, for $X_i = x_i$, where $S_k = 1 - (X_1 + \cdots + X_{k-1})$ is $\beta(n_k, n_{k+1})$.

178. Let $X = (X_1, X_2, X_3)$ be uniformly distributed in the subset of the positive orthant defined by $x_1 \geq 0, x_2 \geq 0, x_3 \geq 0, x_1 + x_2 + x_3 \leq c$. Find, (a) the density of X, (b) the marginal distribution of (X_1, X_2).

179. A bivariate normal distribution has density

$$f(x, y) = c \exp[-x^2 + xy - y^2].$$

(a) Find the constant c and the moments of order 2.
(b) Find the curves of constant density $f(x, y) = c^*$.

180. If (X, Y) is uniform in the triangle $x \geq 0, y \geq 0, x + y \leq 2$, find, (a) the density of (X, Y), (b) the density of X, (c) the conditional density of Y for $X = x$, (d) $E(Y|X = x)$.

181. A die is thrown 12 times.
(a) Find the probability that every face appears twice.
(b) Let X be the number of appearances of 6 and Y the number of appearances of 1. Find the joint distribution of X, Y.
(c) Find $\text{Cov}(X, Y)$.

182.* Show that the most probable value $(n_1^0, n_2^0, \ldots, n_{k+1}^0)$ of the multinomial distribution satisfies the relations

$$np_i - 1 < n_i^0 \le (n + k)p_i, \qquad i = 1, 2, \ldots, k + 1.$$

Hint: Show that

$$p_i n_j^0 \le p_j(n_i^0 + 1) \qquad \text{for } i \ne j, \quad i, j = 1, 2, \ldots, k + 1.$$

183.* *Multiple Poisson distribution.* If the number of trials n is large and the p_i small so that the $np_i = \lambda_i$ are moderate, show that the multinomial distribution can be approximated by the so-called multiple Poisson distribution

$$e^{-(\lambda_1 + \lambda_2 + \cdots + \lambda_k)} \frac{\lambda_1^{n_1} \lambda_2^{n_2} \ldots \lambda_k^{n_k}}{n_1! \, n_2! \ldots n_k!}.$$

184. Given that the density of X and Y is

$$f(x, y) = \frac{2}{(1 + x + y)^3}, \qquad x > 0, \quad y > 0,$$

find, (a) $F(x, y)$, (b) $f_X(x)$, (c) $f_Y(y|X = x)$.

185. Find the density function $f(x, y)$ of the uniform distribution in the circle $x^2 + y^2 \le 1$. Find the marginal distributions of X and Y. Are the variables X and Y independent?

186. The joint density of the variables X, Y, Z is

$$f(x, y, z) = 8xyz, \qquad 0 < x, y, z < 1.$$

Find $P[X < Y < Z]$.

187. For each of the following densities $f(x, y)$, find $F(x, y)$, $F_X(x)$, $F_Y(y)$, $f_X(x)$, $f_Y(y)$, $f_X(x|Y = y)$, $f_Y(y|X = x)$.

(a) $\qquad f(x, y) = 4xy, \qquad 0 < x, \quad y < 1,$

(b) $\qquad f(x, y) = \frac{1}{8}(x^2 - y^2)e^{-x}, \qquad 0 < x < \infty, \quad |y| < x.$

188. The joint probability function $p(x_i, y_j)$ of X, Y is given in the following table:

x \ y	−1	0	1	2	$p(x_j)$
−1	0.1	0.15	0	0.1	0.35
0	0.15	0	0.1	0.2	0.45
1	0.05	0.05	0	0.1	0.20
$q(y_j)$	0.3	0.2	0.1	0.4	1.00

(a) Calculate the regression lines of X on Y and of Y on X.

(b) If it is possible to observe only X, what is the best estimate of Y in terms of X in the sense of the mean square error?

189.* Show that the discrete variables X and Y with joint probability function $p_{ij} = P[X = x_i, Y = y_j]$ are independent if and only if the matrix of probabilities $P = (p_{ij})$ has rank 1.

190. Show that two orthogonal (uncorrelated) variables each taking two values are independent.

CHAPTER 6

Generating Functions. Characteristic Functions

Elements of Theory

1. The probability or factorial moment generating function $P(t)$ of a non-negative integer-valued random variable X is defined by

$$P(t) = E(t^X) = \sum_{k=0}^{\infty} P[X = k]t^k = \sum p_k t^k.$$

It exists at least for $-1 \leq t \leq 1$.

If the factorial moment π_r, say, of order r exists,

$$\pi_r = E[X(X - 1)\ldots(X - r + 1)] = E[(X)_r],$$

then this is given by

$$\pi_r = P^{(r)}(1), \qquad r = 1, 2, \ldots, \tag{6.1}$$

where $P^{(r)}(t)$ denotes the derivative of order r of $P(t)$. Thus we obtain

$$\pi_1 = E(X) = P'(1), \qquad \pi_2 = E(X^2) - E(X) = P''(1),$$

and hence,

$$\text{Var}(X) = P''(1) + P'(1) - [P'(1)]^2. \tag{6.2}$$

2. Distribution of the sum of a random number of random variables. Let

$$S_N = X_1 + X_2 + \cdots + X_N, \tag{6.3}$$

where the random variable $N = 0, 1, \ldots$ and the X_i are completely independent. If the X_i have the same distribution with probability generating function $P_X(t)$, then the probability generating function of S_N

$$P_{S_N}(t) = P_N(P_X(t)). \tag{6.4}$$

Moreover, if $E(X)$ and $E(N)$ exist, then

$$E(S_N) = E(X)E(N). \tag{6.5}$$

3. *Compound Poisson distribution.* This is the distribution of S_N when N has the Poisson distribution. Therefore its probability generating function is

$$P_{S_N}(t) = e^{\lambda(P_X(t)-1)}, \tag{6.6}$$

where $\lambda = E(N)$. If the X_i are Bernoulli random variables, then S_N is Poisson with parameter λp, where $p = E(X_i)$.

4. The moment generating function $M(t)$ of the random variable X is defined by

$$M(t) = E(e^{tX})$$

provided it exists for $-\delta < t < \delta$ and some $\delta > 0$.

Proposition. If $M_X(t)$ exists, then it has derivatives of every order for $|t| < \delta$ ($\delta > 0$) and the moments of every order exist. In fact, we have for every $r = 1, 2, \ldots$

$$E(X^r) = M^{(r)}(0),$$

where $M^{(r)}(t)$ denotes the rth derivative of M. Furthermore, it determines uniquely (characterizes) the distribution of X.

5. *Complex random variable*: $Z = X + iY$ where the X and Y have a joint probability function. Its expected value is defined by

$$E(Z) = E(X) + iE(Y), \qquad i = \sqrt{-1}.$$

6. The characteristic function of a random variable X is defined by

$$\varphi(t) = E(e^{itX}) = E(\cos tX) + iE(\sin tX).$$

It exists for every value of the real parameter t since $|e^{itx}| = 1$.

Properties of a characteristic function $\varphi(t)$
  (i) $\varphi(0) = 1, |\varphi(t)| \leq 1$.
  (ii) $\varphi(-t) = \overline{\varphi(t)}$ where $\bar{a}$ denotes the conjugate of a complex number a.
  (iii) $\varphi(t)$ is uniformly continuous everywhere.
  (iv) If the rth moment $\mu'_r = E(X^r)$ exists, then $\varphi(t)$ has a derivative of order r, $\varphi^{(r)}(t)$, and

$$\mu'_r = i^{-r}\varphi^{(r)}(0),$$

thus $\mu'_1 = E(X) = i^{-1}\varphi'(0), \mu'_2 = -\varphi''(0)$.
  (v) If $\varphi(t)$ has a derivative of order r at $t = 0$, then all moments up to order r exist if r is even and up to order $r - 1$ if r is odd.

(vi) If $E(X^r)$ exists then $\varphi(t)$ has the Mac Laurin expansion

$$\varphi(t) = \sum_{k=0}^{r} \mu'_k \frac{(it)^k}{k!} + O(t^k), \qquad (6.7)$$

where $O(x)$ denotes a quantity such that $O(x)/x \to 0$ as $x \to 0$.

Moreover, the so-called *cumulant generating function* or *second characteristic function*

$$\psi(t) = \log \varphi(t)$$

has the expansion

$$\psi(t) = \sum_{j=0}^{r} \kappa_j \frac{(it)^j}{j!} + O(t^r), \qquad (6.8)$$

where κ_j is called the cumulant of order j of X.

Note: For a proof of (iv)–(vi) we refer, e.g., to Cramér (1946).

7. The characteristic function of a random variable X determines uniquely (characterizes) the distribution of X. In fact, we have the following *inversion formula* of characteristic functions (Fourier transforms).

If $x + h$ and $x - h \, (h > 0)$ are continuity points of the distribution function F of X then

$$F(x + h) - F(x - h) = \lim_{c \to \infty} \frac{1}{\pi} \int_{-c}^{c} \frac{\sinh t}{t} e^{-ixt} \varphi(t) \, dt.$$

For a discrete random variable we have

$$p_k = P[X = x_k] = F(x_k) - F(x_k-) = \lim_{c \to \infty} \frac{1}{2c} \int_{-c}^{c} e^{-ix_k t} \varphi(t) \, dt. \qquad (6.9)$$

In the special case when X is integer valued we have

$$p_k = P[X = k] = \frac{1}{2\pi} \int_{-\pi}^{\pi} e^{-ikt} \varphi(t) \, dt.$$

If the characteristic function $\varphi(t)$ is absolutely integrable in $(-\infty, \infty)$, then the corresponding distribution is continuous with density $f(x)$, given by the inversion formula

$$f(x) = \frac{1}{2\pi} \int_{-\infty}^{\infty} e^{-ixt} \varphi(t) \, dt. \qquad (6.10)$$

8. *Common property of the probability generating function, the moment generating function, and the characteristic function.* This is very useful for the study of the distribution of sums of independent random variables. Let $G(t)$ denote any of these functions. Then, if $S_n = X_1 + X_2 + \cdots + X_n$ where the X_i

are independent we have

$$G_{S_n}(t) = \prod_{i=1}^{n} G_{X_i}(t). \tag{6.11}$$

9. If $P(t)$, $M(t)$, and $\varphi(t)$ exist for a random variable X, then the following relations hold:

$$M_X(t) = P_X(e^t) = \varphi_X(-it),$$

$$\varphi_X(t) = M_X(it) = P_X(e^{it}).$$

10. In addition to characterizing a distribution, a characteristic function facilitates the study of the asymptotic behavior of a sequence of random variables as indicated by the following.

Continuity theorem for characteristic functions (Lévy–Cramér). Let $\{X_n\}$ be a sequence of random variables and let $\{F_n(x)\}$, $\{\varphi_n(t)\}$ be the corresponding sequences of distribution functions and characteristic functions. Then a necessary and sufficient condition that the sequence $F_n(x)$ of distribution functions converges to a distribution function $F(x)$ for every continuity point x of F (weak convergence) is that $\varphi_n(t) \to \varphi(t)$, as $n \to \infty$ for every t and $\varphi(t)$ is continuous at $t = 0$. Whenever this holds the limiting characteristic function $\varphi(t)$ is the characteristic function of the limiting distribution function F.

11. *Infinitely divisible distributions.* The random variable X or its distribution is called *infinitely divisible* if for every integer n, X can be represented as the sum of n independent and identically distributed random variables, i.e., $X = X_1 + \cdots + X_n$.

Hence the characteristic function $\varphi(t)$ of X must satisfy

$$\varphi(t) = [\varphi_n(t)]^n \qquad \text{for every } n,$$

where $\varphi_n(t)$ is a characteristic function.

12. *Generating functions for multivariate distributions.* The definitions of probability generating functions, moment generating functions, and characteristic function(s) of univariate distributions extend easily to multivariate distributions of $(X_1, \ldots, X_n) = \mathbf{X}'$ by replacing t by the row vector $\mathbf{t}' = (t_1, \ldots, t_n)$ and X by $\mathbf{X}$, so that tX is replaced by $\mathbf{t}'\mathbf{X} = t_1 X_1 + \cdots + t_n X_n$.

Thus, for example, the probability generating function of X and Y with joint probability function

$$p_{jk} = P[X = j, Y = k], \qquad j, k = 0, 1, \ldots,$$

is defined by

$$P(t_1, t_2) = E(t_1^X t_2^Y) = \sum_{j,k} p_{jk} t_1^j t_2^k,$$

and the characteristic function of $\mathbf{X} = (X_1, \ldots, X_n)'$ by

$$\varphi_{\mathbf{X}}(\mathbf{t}) = \varphi_{\mathbf{X}}(t_1, \ldots, t_n) = E(e^{i\mathbf{t}'\mathbf{X}}) = Ee^{i(t_1 X_1 + \cdots + t_n X_n)}.$$

It can easily be shown that:

(i) $P_X(t_1) = P(t_1, 1), P_Y(t_2) = P(1, t_2)$.

(ii) $P_{X+Y}(t) = P(t, t)$.

(iii) X and Y are independent if and only if

$$P(t_1, t_2) = P(t_1, 1)P(1, t_2) = P_X(t_1)P_Y(t_2) \qquad \text{for every } (t_1, t_2),$$

$$\varphi(t_1, t_2) = \varphi(t_1, 0)\varphi(0, t_2) = \varphi_X(t_1)\varphi_Y(t_2) \qquad \text{for every } (t_1, t_2).$$

(iv) The mixed moment $\mu'_{jk} = E(X^j Y^k)$ is given by

$$\mu'_{jk} = \frac{\partial^{j+k}\varphi(0, 0)}{\partial t_1^j \, \partial t_2^k} i^{-(j+k)}.$$

By analogy to the inversion formula for univariate distributions, we have the following:

If the vertices of the generalized rectangle (or n-cell) $x_i - h_i \le X_i \le x_i + h_i$ $(i = 1, \ldots, n)$ $(h_i > 0)$ are continuity points of the distribution function $F(x_1, \ldots, x_n)$, then

$$P[x_i - h_i < X_i \le x_i + h_i, i = 1, \ldots, n]$$

$$= \lim_{c \to \infty} \frac{1}{\pi^n} \int_{-c}^{c} \cdots \int_{-c}^{c} \varphi(t_1, \ldots, t_n) \prod_{i=1}^{n} \frac{\sin h_i t_i}{t_i} \exp(-ix_i t_i) \, dt_i.$$

In the special case of an absolutely integrable $\varphi(t_1, \ldots, t_n)$, the distribution is continuous with density

$$f(x_1, \ldots, x_n) = \frac{1}{(2\pi)^n} \int_{-\infty}^{\infty} \cdots \int_{-\infty}^{\infty} e^{-i(x_1 t_1 + \cdots + x_n t_n)} \varphi(t_1, \ldots, t_n) \, dt_1 \ldots dt_n.$$

The continuity theorem also holds, i.e.,

$$F_\nu(x_1, \ldots, x_n) \to F(t_1, \ldots, t_n) \Leftrightarrow \varphi_\nu(t_1, \ldots, t_n) \to \varphi(t_1, \ldots, t_n) \quad \text{as } \nu \to \infty$$

when φ is continuous at $(0, \ldots, 0)$.

Exercises

191. Find the probability generating function and hence the mean and variance of the following distributions: (a) binomial, (b) Poisson, (c) geometric, (d) Pascal. Moreover, deduce the corresponding moment generating functions and characteristic functions.

192.* Let $X_1 = X_2 = X$ when X has the Cauchy density

$$f(x) = \frac{1}{\pi} \cdot \frac{1}{1 + x^2}, \qquad -\infty < x < \infty.$$

Prove that

$$\varphi_{X_1+X_2}(t) = \varphi_{X_1}(t)\varphi_{X_2}(t),$$

while X_1, X_2 are not independent.

193. From the probability generating function $P_X(t)$ of $X = 0, 1, 2, \ldots$, find the probability generating function of $Y = 3X + 2$.

194. Using the probability generating function prove that the random variable S_N of (6.3) has variance given by $\text{Var}(S_N) = E(N)\text{Var}(X) + \text{Var}(N)E^2(X)$.

195. A woman continues to have children until she has a boy! Suppose that the probability of having a blond child is p; find the probability that the woman will have k blond children. (Assume that the color is independent of sex.)

196. The number N of visitors to a shop has the Poisson distribution. On average, 150 women visit the shop every day. Of these, $100p_1\%$ buy mini, $100p_2\%$ buy midi, $100p_3\%$ buy maxi, and the remaining buy nothing. (No woman buys more than one dress!!) Let X_1, X_2, X_3 be the corresponding numbers of mini, midi, and maxi that are bought in a day. Find the joint probability function of X_1, X_2, X_3. Are they independent?

197.* *Equal characteristic functions in a finite interval do not necessarily define the distribution uniquely.* Let X, Y be independent with densities

$$f_X(u) = \frac{a + b - a\cos(u/a) - b\cos(u/b)}{2\pi u^2},$$

$$-\infty < u < +\infty.$$

$$f_Y(u) = \frac{c - c\cos(u/c)}{\pi u^2},$$

Prove that $\varphi_X(t) = \varphi_Y(t)$ for $|t| \leq \min\{1/a, 1/b, 1/c\}$, $2c = a + b$.

198.* Let $X_1, X_2, \ldots, X_n$ be independent random variables having the same normal distribution with mean μ and variance 1. We set

$$S_n = \sum_{k=1}^{n} X_k^2.$$

Moreover, let the random variable $T = aY_v$ where a, v are suitable constants, and Y_v follows the χ^2 distribution with v degrees of freedom. Using characteristic functions, find the constants a and v so that the random variables S_n and T have the same means and the same variances.

199. Prove that S_N of (6.3) has the characteristic function

$$\varphi_{S_N}(t) = P_N(\varphi_{X_i}(t)).$$

200. Prove that $\varphi(t) = \exp[\lambda(e^{-|t|} - 1)]$ is the characteristic function of a Poisson distribution compounded with (generalized by) a Cauchy distribution. Then, using the inversion formula, find its density.

201.* Let Z_n be a random variable following the binomial distribution with parameters n and $p = \lambda/n$ $(\lambda > 0)$. Moreover, let the random variable Z have a Poisson distribution with parameter λ. Prove that $\lim_{n \to \infty} \varphi_{Z_n}(t) = \varphi_Z(t)$ $(t \in R)$, where $\varphi_{Z_n}(t)$ and $\varphi_Z(t)$ are the characteristic functions of Z_n and Z, respectively. Then deduce the approximation of the binomial by the Poisson.

202. Calculate the characteristic function of a Gamma distribution with density

$$f(x) = \frac{\lambda^s}{\Gamma(s)} e^{-\lambda x} x^{s-1}, \qquad x > 0 \quad (s > 0).$$

Then deduce the characteristic function of χ_n^2.

203. At the nth toss of a coin a player receives or pays 2^{-n} of a drachma depending on whether heads or tails appears. Let Y_n be the gain of the player after n tosses. Show that the distribution of Y_n as $n \to \infty$ tends to the uniform distribution in the interval $(-1, 1)$. First find the characteristic function of Y_n.

204. X has distribution function $F(x)$ which is a mixture of two distribution functions F_1, F_2 as follows:

$$F(x) = \lambda F_1(x) + (1 - \lambda) F_2(x), \qquad \lambda > 0.$$

If F_1 is $N(\mu_1, \sigma_1^2)$ and F_2 is $N(\mu_2, \sigma_2^2)$ find the characteristic function of F and then the $E(X)$ and $\mathrm{Var}(X)$.

205. Find the second characteristic function (cumulant generating function) $\psi(t)$ of the exponential with density

$$f(x) = \theta e^{-\theta x}, \qquad x > 0,$$

and prove that the cumulant of order r κ_r is given by

$$\kappa_r = \frac{(r - 1)!}{\theta^r}, \qquad r = 1, 2, \dots.$$

206. (*Continuation*). Let Y be a discrete random variable defined as follows. For $h > 0$ and $k = 0, 1, 2, \dots$,

$$P[Y = (2k + 1)h/2] = P[kh \le X \le (k + 1)h].$$

Find the probability generating function of Y and then show that $E(X) > E(Y)$, $\mathrm{Var}(X) < \mathrm{Var}(Y)$.

207.* Suppose that the density $f(x, y) = g(x^2 + y^2)$, i.e., it has circular symmetry about the origin. Show that the characteristic function of f, $\varphi(t, u)$, is a function of $t^2 + u^2$ (only). Hence show that the only distribution of independent random variables with the above property is the normal. (Consider an orthogonal transformation of t, u.)

208. Consider two independent random samples $X_1, X_2, \dots, X_n$ and

$Y_1, Y_2, \ldots, Y_n$ from the Laplace distribution

$$f(x) = \tfrac{1}{2}e^{-|x|}, \qquad -\infty < x < \infty.$$

Show that the means of the differences $X_i - Y_i$ and $X_i + Y_i$ have the same distribution but are not independent. Are they orthogonal?

209. If $\varphi(t)$ is a characteristic function show that, (a) $|\varphi(t)|^2$, (b) $e^{\lambda(\varphi(t)-1)}$ are also characteristic functions.

210.* Calculate the probability generating function of the multinomial distribution and then show that

$$E(X_j) = np_j, \qquad \mathrm{Var}(X_j) = np_j(1 - p_j), \qquad \mathrm{Cov}(X_j, X_k) = -np_jp_k.$$

211. (*Continuation*). If, as $n \to \infty$, $np_j \to \lambda_j$ $(j = 1, 2, \ldots, k + 1)$, then the multinomial distribution tends to the k-dimensional Poisson $Y = (Y_1, Y_2, \ldots, Y_k)$ where the Y_i are independent Poisson, Y_j with parameter λ_j (cf. Exercise 183).

212.* Calculate the characteristic function of the negative multinomial distribution (5.9) and hence
(a) Verify that $E(X_j) = s\theta_j$, $\mathrm{Var}(X_j) = s\theta_j(1 + \theta_j)$, $\mathrm{Cov}(X_j, X_k) = s\theta_j\theta_k$.
(b) Show that if the conditional distributions of the X_j given $V = v$ are independent Poisson with parameters $v\lambda_j$, respectively $(j = 1, 2, \ldots, r)$ and V has the Gamma distribution with density

$$f(v) = \frac{\lambda^s}{\Gamma(s)} e^{-\lambda v} v^{s-1}, \qquad s > 0,$$

then $X = (X_1, X_2, \ldots, X_k)$ has the above negative multinomial distribution with parameters $\theta_j = \lambda_j/\lambda$.

213. X, Y have a distribution of the continuous type with characteristic function $\varphi(t_1, t_2)$. Show that the conditional characteristic function of X given $Y = y$ is

$$\varphi_X(t_1 \mid Y = y) = \frac{\displaystyle\int_{-\infty}^{\infty} e^{-it_2 y} \varphi(t_1, t_2)\, dt_2}{\displaystyle\int_{-\infty}^{\infty} e^{-it_2 y} \varphi(0, t_2)\, dt_2}.$$

214. A random sample $X_1, X_2, \ldots, X_{n_1+n_2-n}$ on a continuous random variable X with $-\infty < X < \infty$, is divided at random into three subsets of $(n_1 - n)$, $(n_2 - n)$, and n observations $(n_1 > n, n_2 > n)$. Let $S_1, S_2,$ and S_3 denote the corresponding sums. Find the characteristic function of

$$Y = S_1 + S_3, \qquad Z = S_2 + S_3 \qquad \text{and} \qquad E(Y \mid Z = z).$$

215. Show that the following distributions are infinitely divisible: (a) Pascal, (b) Cauchy, (c) Laplace, (d) Gamma.

CHAPTER 7

Distribution of Functions of Random Variables

Elements of Theory

In finding the distribution of a function of a scalar or vector random variable the following formulas turn out to be very useful.

1. *Distribution of monotone functions.* Let X be a continuous random variable with distribution function F_X and $y = g(x)$ a differentiable and monotone (increasing or decreasing) function with inverse function $x = g^*(y)$ (i.e., $g'(x)$ is everywhere either > 0 or < 0). Then the random variable $Y = g(X)$ is also continuous with distribution function F_Y given by

$$F_Y(y) = \begin{cases} F_X(g^*(y)), & g'(x) > 0, \\ 1 - F_X(g^*(y)), & g'(x) < 0, \end{cases}$$

and density function

$$f_Y(y) = f_X(g^*(y)) \left| \frac{dg^*}{dy} \right| = f_X(g^*(y)) \frac{1}{|g'(x)|}. \tag{7.1}$$

2. Under the preceding assumptions, except that the equation

$$y = g(x)$$

has roots $x_1, \ldots, x_n, \ldots$, i.e., for given y we have

$$y = g(x_1) = g(x_2) = \cdots = g(x_n) = \cdots,$$

and the derivatives

$$g'(x_i) \neq 0, \qquad i = 1, 2, \ldots,$$

$Y = g(X)$ is also continuous with density

$$f_Y(y) = \frac{f_X(x_1)}{|g'(x_1)|} + \cdots + \frac{f_X(x_n)}{|g'(x_n)|} + \cdots. \tag{7.2}$$

3. *Functions of two random variables.* An interesting case is when we have a function of two continuous random variables X and Y with joint density $f(x, y)$.

(a) *Sum of two random variables.* The distribution function $F_Z(z)$ of the sum

$$Z = X + Y$$

is given by

$$
\begin{aligned}
F_Z(z) &= \iint_{x+y \leq z} f(x, y)\, dx\, dy = \int_{-\infty}^{\infty} dy \int_{-\infty}^{z-y} f(x, y)\, dx \\
&= \int_{-\infty}^{\infty} dx \int_{-\infty}^{z-x} f(x, y)\, dy = \int_{-\infty}^{\infty} dx \int_{-\infty}^{z} f(x, y - x)\, dy \\
&= \int_{-\infty}^{\infty} dy \int_{-\infty}^{z} f(x - y, y)\, dx.
\end{aligned}
$$

Differentiating with respect to z we have the density of Z

$$f_Z(z) = \int_{-\infty}^{\infty} f(z - y, y)\, dy = \int_{-\infty}^{\infty} f(x, z - x)\, dx.$$

In the special case of independence of X and Y, the density $f_Z(z)$ of the sum is called the *convolution* of the density functions f_X and f_Y, and we have

$$f_Z(z) = \int_{-\infty}^{\infty} f_X(z - y) f_Y(y)\, dy = \int_{-\infty}^{\infty} f_X(x) f_Y(z - x)\, dx. \tag{7.3}$$

The distribution function F_Z is given by

$$F_Z(z) = \int_{-\infty}^{\infty} F_X(z - y) f_Y(y)\, dy = \int_{-\infty}^{\infty} F_Y(z - x) f_X(x)\, dx. \tag{7.4}$$

Note. If X and Y are discrete, the integrals are replaced by sums.

(b) *Distribution of the quotient of two continuous random variables.* The distribution function F_Z of the ratio

$$Z = \frac{X}{Y}$$

is given by

$$F_Z(z) = \int_{0}^{\infty} dy \int_{-\infty}^{yz} f(x, y)\, dx + \int_{-\infty}^{0} dy \int_{zy}^{\infty} f(x, y)\, dx, \tag{7.5}$$

and the density by

$$f_Z(z) = \int_0^\infty yf(yz, y)\, dy - \int_{-\infty}^0 yf(yz, y)\, dy. \tag{7.6}$$

In the special case of independence of X and Y, (7.5) and (7.6) become

$$F_Z(z) = \int_0^\infty F_X(zy) f_Y(y)\, dy + \int_{-\infty}^0 [1 - F_X(zy)] f_Y(y)\, dy,$$

$$f_Z(z) = \int_{-\infty}^\infty |y| f_X(zy) f_Y(y)\, dy. \tag{7.7}$$

4. *Functions of several random variables.* This is the general case in which, given the joint density function $f(x_1, \ldots, x_n)$ of the random variables $X_1, \ldots, X_n$, we want the distribution of

$$Y_j = g_j(X_1, \ldots, X_n), \qquad j = 1, \ldots, m. \tag{7.8}$$

Then the joint distribution function $G(y_1, \ldots, y_m)$ of $Y_1, \ldots, Y_m$ can be obtained from

$$G(y_1, \ldots, y_m) = \int \cdots_D \int f(x_1, \ldots, x_n)\, dx_1, \ldots, dx_n,$$

where the region of integration D is defined by the relation

$$D = \{(x_1, \ldots, x_n): g_j(x_1, \ldots, x_n) \le y_j, j = 1, \ldots, m\}.$$

For discrete random variables the distribution is obtained by replacing the multiple integral by a multiple sum over D. Of special interest is the case $m = n$ for which the problem, under certain conditions, has a specific solution.

If the functions g_j of (7.8) ($j = 1, \ldots, n$) have continuous partial derivatives of the first order so that the Jacobian J of the transformation (7.8) is $\neq 0$, i.e.,

$$J(x_1, \ldots, x_n) = \begin{vmatrix} \dfrac{\partial g_1}{\partial x_1} & \cdots & \dfrac{\partial g_1}{\partial x_n} \\ \vdots & & \\ \dfrac{\partial g_n}{\partial x_1} & \cdots & \dfrac{\partial g_n}{\partial x_n} \end{vmatrix} \neq 0, \tag{7.9}$$

then there exists an inverse transformation

$$x_i = g_i^*(y_1, \ldots, y_n), \qquad i = 1, \ldots, n, \tag{7.10}$$

and the random variables $Y_1, \ldots, Y_n$ are continuous with joint density (cf. (7.1))

$$f^*(y_1, \ldots, y_n) = f(g_1^*(y_1, \ldots, y_n), \ldots, g_n^*(y_1, \ldots, y_n))|J(y_1, \ldots, y_n)|, \tag{7.11}$$

where $J(y_1, \ldots, y_n)$ denotes the Jacobian of the inverse transformation (7.10), i.e.,

$$J(y_1, \ldots, y_n) = \left| \left(\frac{\partial g_i^*}{\partial y_j} \right) \right| = [J(x_1, \ldots, x_n)]^{-1}.$$

For example, in the case of linear transformations

$$y_i = \sum_{j=1}^{n} a_{ij} x_j, \qquad i = 1, \ldots, n,$$

the Jacobian equals the determinant of $A = (a_{ij})$,

$$J(x_1, \ldots, x_n) = |A|.$$

When A is nonsingular, i.e., $|A| \neq 0$, the density of the random vector

$$Y = AX,$$

where $Y = (Y_1, \ldots, Y_n)'$, $X = (X_1, \ldots, X_n)'$ is given by

$$f^*(y_1, \ldots, y_n) = f(A^{-1}y)|A|^{-1}. \tag{7.12}$$

Exercises

216. The radius of a circle is approximately measured so that it has the uniform distribution in the interval (a, b). Find the distribution:
(a) of the length of the circumference of the circle;
(b) of the area of the circle.

217. Suppose that X and Y are independent random variables with the same exponential density

$$f(x) = \theta e^{-\theta x}, \qquad x > 0.$$

Show that the sum $X + Y$ and the ratio X/Y are independent.

218. Suppose that X and Y are independent uniform variables in $(0, 1)$. Find the probability that the roots of $\lambda^2 + 2X\lambda + Y = 0$ are real.

219. The angle φ at which a projectile is fired with initial velocity v follows the uniform distribution in the interval $(0, \pi/2)$. Find the distribution of the horizontal distance d between the firing point and the point at which the projectile falls.

220.* Suppose that X and Y are normal variables with mean 0, variance σ^2, and correlation coefficient ρ.
(a) Find the distribution of X/Y.
(b) Using the above result prove that

$$P[X < 0, Y > 0] = \frac{1}{2} P[XY < 0] = \frac{1}{4} - \frac{\arcsin \rho}{2\pi}.$$

Verify this by integrating the density of (X, Y).

221. Let $A = (X, Y)$ be the point of impact of a shot on a vertical target with center at origin O and suppose the distribution of A is that of Exercise 220 with $\rho = 0$ (called circular normal). Show that the distance $R = (OA)$ between the point of impact and the centre of the target and the angle of OA with the horizontal axis (i.e., the polar coordinates of A) are independent variables. Find the distribution of R^2.

222. If X has the uniform distribution on the interval $(0, 1)$, then:
(a) Prove that $Y = aX + b$ is also uniformly distributed (a and b constants).
(b) Find the distribution of $Y = AX^2 + BX + C$ (A, B, C constants).

223. If X_1, X_2 are independent and uniform on the interval $(0, 1)$, find the densities of: (a) $X_1 + X_2$, (b) $X_1 - X_2$, (c) $|X_1 - X_2|$, (d) X_1/X_2.

224. Two friends A and B agree to meet between 12 (noon) and 1 p.m. at a restaurant. Supposing that they arrive at random between 12 and 1 p.m. independently of each other and the lunch lasts 30 minutes, what is the probability that they meet in the restaurant? Let T be the instant of their meeting. What is the conditional distribution of T, (a) given that they meet, (b) given that they meet and A arrives first?

225. Let X and Y be independent with densities

$$f_X(x) = \frac{1}{\pi} \frac{1}{\sqrt{1 - x^2}}, \quad |x| < 1, \qquad f_Y(y) = \frac{y}{\sigma^2} e^{-y/2\sigma^2}.$$

Show that XY is $N(0, \sigma^2)$.

226.* Given n independent random numbers $X_1, X_2, \ldots, X_n$ from $(0, 1)$, called pseudorandom numbers, show that:
(a) $Y = -\sum_{i=1}^{n} 2 \log X_1$ has the χ^2 distribution with $2n$ degrees of freedom;
(b) the variables

$$\xi = -\sqrt{-2 \log X_1} \, \cos(2\pi X_2),$$
$$\eta = \sqrt{-2 \log X_1} \, \sin(2\pi X_2),$$

are independent $N(0, 1)$.
Thus (a) generates samples of χ^2 distributions with an even number of degrees of freedom, while (b) generates a pair of independent normal variables.

227.* Let X be a continuous random variable X with distribution function $F(\cdot)$; consider the random variable

$$Y = F(X) = \int_{-\infty}^{X} f(u) \, du.$$

This transformation is called the *probability integral transformation*.
Show that the distribution of Y is uniform in the interval $(0, 1)$, that is,

$$F_Y(y) = y, 0 < y < 1.$$

Given a random sample $X_1, X_2, \ldots, X_n$ from the uniform distribution, show that the solutions $y_1, y_2, \ldots, y_n$ of $y_1 = F^{-1}(x_1), y_2 = F^{-1}(x_2), \ldots, y_n = F^{-1}(x_n)$ form a random sample from the F distribution.

228. (*Continuation*). Suppose the height X of men has the normal distribution $N(167 \text{ cm}, 9 \text{ cm}^2)$ and a random sample of 10 men is taken with heights $X_1, X_2, \ldots, X_{10}$. What is probability that the (random) interval $\delta = [\min(X_1, \ldots, X_{10}), \max(X_1, \ldots, X_{10})]$ covers at least 95% of the population of heights? That is, the $P\{P[X \in \delta] \geq 0.95\}$ is to be calculated; note that the $P[X \in \delta]$, depending on the random interval δ, is a random variable.

229. X is called a lognormal variable, if the $\log X = Y$ has a normal distribution $N(\mu, \sigma^2)$.
   (i) Find, (a) the density of X, (b) $E(X)$ and $\text{Var}(X)$.
   (ii) If the X_i are independent lognormal random variables, their product $X_1 X_2 \ldots X_n$ is also lognormal.

230. Show that if $F(x, y)$ is the distribution function of X and Y and

$$Z = \max(X, Y), \qquad W = \min(X, Y),$$

then

$$F_Z(z) = F(z, z), \qquad F_W(w) = F_X(w) + F_Y(w) - F(w, w).$$

If $F(x, y)$ is continuous find the densities of Z and W.

231. (*Continuation*). If X and Y are independent $N(0, 1)$, show that

$$E\{\max(X, Y)\} = \frac{1}{\sqrt{\pi}}.$$

232.* Let $X = (X_1, X_2, \ldots, X_k)$ have the Dirichlet distribution. Find the distribution, (a) of $Y_k = X_1 + X_2 + \cdots + X_k$, and (b) of $Y_i = X_1 + X_2 + \cdots + X_i$ $(i = 1, 2, \ldots, k)$.

233.* (*Continuation*). If X_i $(i = 1, 2, \ldots, k + 1)$ are independent Gamma variables with parameters $n_1, \ldots, n_{k+1}$ and λ, respectively, show that the

$$Y_i = \frac{X_i}{X_1 + \cdots + X_{k+1}}, \qquad i = 1, 2, \ldots, k,$$

have the k-dimensional Dirichlet distribution (5.11).

234. Let X and Y be independent Poisson variables with parameters λ and μ. Show that:
   (a) the sum $X + Y$ is also Poisson (use the convolution formula, cf. (7.3));
   (b) the conditional distribution of X given that $X + Y$ is binomial.

235. Let X and Y be independent binomial with parameters N, p and M, p, respectively. Show that:
   (a) the sum of $X + Y$ is binomial (use the convolution formula);
   (b) the conditional distribution of X given $X + Y$ is hypergeometric, independent of p.

236.* Show that the mean

$$\bar{X} = \frac{1}{n} \sum_{i=1}^{n} X_i$$

and the variance

$$s^2 = \frac{1}{n-1} \sum_{i=1}^{n} (X_i - \bar{X})^2$$

of the random sample $X_1, X_2, \ldots, X_n$ from $N(\mu, \sigma^2)$ are independent. Hence deduce the distribution of $(n-1)s^2$.

Hint: (a) Use any orthogonal transformation $\mathbf{Y} = H\mathbf{X}$ such that

$$Y_n = \sqrt{n}\bar{X}, \qquad \mathbf{Y} = (Y_1, \ldots, Y_n)', \qquad \mathbf{X} = (X_1, \ldots, X_n)'.$$

(b) $\mathrm{Cov}(\bar{X}, X_i - \bar{X}) = 0$.
Show that $\mathrm{Cov}(\bar{X}, s^2) = \mu_3/n$ for samples from nonnormal populations, where μ_3 is the population central moment of order 3.

237.* *Test for the Behrens–Fisher problem.* To test whether two hetero-scedastic normal populations have the same means one uses the statistic

$$t = [m(m-1)]^{1/2}(\bar{x} - \bar{y}) \bigg/ \left[\sum_{j=1}^{m} (u_j - \bar{u})^2 \right]^{1/2},$$

where $\bar{x}, \bar{y}$ are the sample means of the independent samples from $N(\mu_1, \sigma_1^2)$ and from $N(\mu_2, \sigma_2^2)$, respectively, $m \le n$ and $u_j = x_j - (\sqrt{m/n})y_j$ ($j = 1, 2, \ldots, m$). Show that t has Student's distribution with $m - 1$ degrees of freedom when $\mu_1 = \mu_2 = \mu$.

238.* Find the joint density of the ordered statistics $X_{(1)} < X_{(2)} < \cdots < X_{(n)}$ from a distribution of the continuous type.

239. If $X_1, X_2, \ldots, X_n$ is a random sample from the continuous distribution F. Find:
(a) the joint density function of the random variables

$$Y = F\left(\min_{1 \le i \le n} X_i \right), \qquad Z = F\left(\max_{1 \le i \le n} X_i \right);$$

(b) $E(R) = E(Z - Y)$ where R is the range of the sample.

240. Let $X_{(1)} < X_{(2)} < \cdots < X_{(n)}$ be the ordered sample from a continuous distribution $F(x)$. Show that

$$E[X_{(k+1)} - X_{(k)}] = \binom{n}{k} \int_{-\infty}^{+\infty} F^{n-k}(x)[1 - F(x)]^k \, dx.$$

Hint: Use the joint density of $X_{(k)}, X_{(k+1)}$.

241.* A random sample of size n is taken from a uniform distribution in the interval $(0, 1)$. Find, (a) the mean $E[X_{(k)}]$, (b) the distribution of $R = X_{(n)} - X_{(1)}$.

242. If $y_1 < y_2 < y_3$ are the ordered observations of a random sample of size 3 from a distribution with density

$$f(x) = \begin{cases} 1, & \theta - \frac{1}{2} < x < \theta + \frac{1}{2}, \\ 0, & \text{elsewhere.} \end{cases}$$

For $\alpha = \beta = 0.4$ show that $P[\theta - \alpha < y_2 < \theta + \beta] = 0.944$.

243. Let $Y_1 < Y_2 < Y_3$ be ordered observations of a random sample of size 3 from a distribution with density

$$f(x) = \begin{cases} 2x; & 0 < x < 1, \\ 0, & \text{elsehwere.} \end{cases}$$

Show that $Z_1 = Y_1/Y_2$, $Z_2 = Y_2/Y_3$, and $Z_3 = Y_3$ are completely independent.

244. If X_1, X_2 are independent β variables with densities $\beta(n_1, n_2)$ and $\beta(n_1 + 1/2, n_2)$, respectively, where for $n_1 > 0, n_2 > 0$,

$$\beta(m, n) = \frac{1}{B(m, n)} x^{m-1}(1 - x)^{n-1}, \qquad 0 < x < 1.$$

Show that $Y = \sqrt{X_1 X_2}$ has also the β distribution $\beta(2n_1, 2n_2)$.

245. The random pair (X, Y) has density

$$f(x, y) = \frac{1}{\Gamma(m)\Gamma(n)} x^{m-1}(y - x)^{n-1}e^{-y}, \qquad 0 < x < y < \infty.$$

Find the distribution of $Z = Y - X$. Are Z and X independent? (cf. Exercise 174).

246. Show that $Z = 2\sqrt{XY}$, where X and Y are two independent Γ variables with parameters (λ_1, n) and $(\lambda_2, n + 1/2)$, respectively, has the Γ distribution with parameters $(\lambda, 2n)$ where $\lambda = \sqrt{\lambda_1 \lambda_2}$.

247. For the pair (X, Y) of Exercise 245 show that $W = X/Y$ has the density $\beta(m, n)$ of Exercise 244.

248. If X and Y are independent $N(0, \sigma^2)$, show that

$$Z = \frac{XY}{\sqrt{X^2 + Y^2}}, \qquad W = \frac{X^2 - Y^2}{\sqrt{X^2 + Y^2}},$$

are also independent normal.

249.* Let X have an n-variate normal distribution $N(\mu, \Sigma)$. Find the density of $Y = AX$ where A is a nonsingular matrix.

250. The points of impact of two players A_1, A_2 on a vertical target follow the circular normal distribution with mean the center of the target and variances σ_1^2 and σ_2^2, respectively. Find the probability that the impact point of A_1 is closer to the center than that of A_2.

CHAPTER 8

Limit Theorems. Laws of Large Numbers. Central Limit Theorems

Elements of Theory

1. *Convergence of a sequence of random variables.* A stochastic sequence, i.e., a sequence $\{X_n\}$ of random variables defined on a probability space $\{\Omega, B, P\}$ may converge in several ways.

(a) *In probability or stochastically or weakly* to a random variable X denoted by

$$X_n \xrightarrow[n \to \infty]{P} X$$

if for every $\varepsilon > 0$

$$\lim_{n \to \infty} P[|X_n - X| > \varepsilon] = 0. \tag{8.1}$$

(b) *With probability one or almost surely (a.s.) or strongly* to a random variable X denoted by

$$X_n \xrightarrow{\text{a.s.}} X$$

if

$$P\left[\lim_{n \to \infty} X_n = X\right] = 1, \tag{8.2}$$

or, equivalently, if for every $\varepsilon > 0$

$$\lim_{N \to \infty} P\left[\sup_{n \geq N} |X_n - X| > \varepsilon\right] = 0. \tag{8.3}$$

(c) *In quadratic mean* to a random variable X, denoted by

$$X_n \xrightarrow{\text{i.m.}} X$$

if

$$\lim_{n \to \infty} (X_n - X)^2 = 0$$

(d) In law or distribution to X, denoted by

$$X_n \xrightarrow{L} X$$

if at every continuity point x of the distribution function F of X

$$\lim_{n \to \infty} F_n(x) = F(x),$$

where F_n denotes the distribution function of X_n.

2. Relations between the modes of convergence. For every constant c the following hold:

(i) $X_n \xrightarrow{P} X \Rightarrow X_n \xrightarrow{L} X$;   (ii) $X_n \xrightarrow{P} c \Leftrightarrow X_n \xrightarrow{L} c$;

(iii) $X_n \xrightarrow{\text{i.m.}} X \Rightarrow X_n \xrightarrow{P} X$;   (iv) $X_n \xrightarrow{\text{a.s.}} X \Rightarrow X_n \xrightarrow{P} X$;   (8.4)

(v) $X_n \xrightarrow{\text{i.m.}} c$ and $\displaystyle\sum_{n=1}^{\infty} E(X_n - c)^2 < \infty \Rightarrow X_n \xrightarrow{\text{a.s.}} c$.

3. *Laws of large numbers* (LLN). They refer to the weak or strong convergence of a sample mean $\bar{X}_n = (1/n) \sum_{i=1}^{n} X_i$ to a corresponding population (distribution) mean μ. Thus we have the

Weak LLN (WLLN):   If $\bar{X}_n \xrightarrow{P} \mu$;

Strong LLN (SLLN):   If $\bar{X}_n \xrightarrow{\text{a.s.}} \mu$.

The basic tool in the proof of the WLLN is the so-called

Chebyshev–Bienaymé inequality. If $E(X) = \mu$, $\text{Var}(X) = \sigma^2$, then for every constant c or λ ($\lambda > 1$)

$$P[|X - \mu| \geq c] \leq \frac{\sigma^2}{c^2} \quad \text{or} \quad P[|X - \mu| \geq \lambda\sigma] \leq \frac{1}{\lambda^2}. \quad (8.5)$$

This follows from

Markov's inequality. If $Y \geq 0$, i.e., $P[Y \geq 0] = 1$ and $E(Y) < \infty$, then for every constant $c > 0$

$$P[Y \geq c] \leq \frac{E(Y)}{c}. \quad (8.6)$$

The Chebyshev–Markov WLLN. If the random variables $X_1, \ldots, X_n, \ldots$ are independent and $E(X_i) = \mu_i$, $\text{Var}(X_i) = \sigma_i^2$ with

$$\text{Var}(\bar{X}_n) \to 0 \quad \text{as} \quad n \to \infty,$$

then

$$\bar{X}_n - \bar{\mu}_n \xrightarrow{P} 0 \qquad \text{where } \bar{\mu}_n = \frac{1}{n}\sum_{i=1}^{n}\mu_i; \tag{8.7}$$

as a consequence we have the WLLN for a random sample (i.e., independent and identically distributed random variables) from a population with mean μ and variance σ^2, i.e.,

$$\bar{X}_n \xrightarrow{P} \mu. \tag{8.8}$$

When the X_i are Bernoulli variables then (8.8) gives, the Bernoulli WLLN, i.e., the proportion $\hat{p}_n = x/n = \bar{X}_n$ of successes $\xrightarrow{P} p$, the probability of success in a trial (see (2.5)).

Khinchine's WLLN. For a random sample $X_1, \ldots, X_n, \ldots$ from a population with mean $\mu = E(X_i)$ $(i = 1, \ldots,)$ the WLLN holds. This follows by the continuity theorem for characteristic function(s) (see Chapter 6) and relation (ii) of (8.4) since

$$\varphi_{\bar{X}_n}(t) \xrightarrow[n\to\infty]{} \varphi_\mu(t) = e^{i\mu t}.$$

The classical Poisson LLN: If the X_i are Bernoulli with

$$E(X_i) = p_i, \qquad i = 1, \ldots,$$

then

$$\bar{X}_n - \bar{p}_n \xrightarrow{P} 0 \qquad \text{where } \bar{p}_n = \frac{1}{n}\sum_{i=1}^{n} p_i,$$

is a special case of (8.7).

Strong laws of large numbers (SLLN). E. Borel (1909) first showed the SLLN for Bernoulli cases (as above)

$$X_n \xrightarrow{\text{a.s.}} p.$$

This means that for every $\varepsilon > 0$ and $\delta > 0$ there exists an integer $N = N(\varepsilon, \delta)$ such that

$$P[|X_n - p| < \varepsilon \text{ for every } n \geq N] \geq 1 - \delta.$$

Later Kolmogorov showed that under the conditions of (8.7)

$$X_n - \bar{\mu}_n \xrightarrow{\text{a.s.}} 0.$$

Moreover for a random sample

$$\bar{X}_n \xrightarrow{\text{a.s.}} \mu \iff E(X_i) = \mu, \tag{8.8}*$$

i.e., the SLLN holds if and only if the population mean exists.

The proofs of these (see, e.g., Gnedenko (1962)) rest on the following inequalities.

Kolmogorov's inequality. Let $X_1, \ldots, X_n$ be independent and let $S_k = X_1 + \cdots + X_k$ with $E(S_k) = m_k$, $\text{Var}(S_k) = s_k^2$ $(k = 1, 2, \ldots)$. Then for every $\varepsilon > 0$

$$P[|S_k - m_k| < \varepsilon s_n, k = 1, 2, \ldots, n] \geq 1 - \frac{1}{\varepsilon^2}. \qquad (8.9)$$

The Hájek–Renyi inequality. Under the conditions of (8.9), for every non-decreasing sequence of positive constants c_n and any integers $m, n, 0 < m < n$,

$$P\left[\max_{m \leq k \leq n} c_k |S_k - m_k| \geq \varepsilon \right] \leq \frac{1}{\varepsilon^2}\left(c_m^2 s_m^2 + \sum_{k=m+1}^{n} c_k^2 \text{Var}(X_k) \right). \qquad (8.10)$$

Central limit theorems (CLT). They refer to the convergence of sums of random variables (or equivalently of sample means) in law. The limiting law is usually the normal. The proofs of CLT's, as a rule, rely on the continuity theorem for characteristic functions (see Chapter 6):

$$X_n \xrightarrow{L} X \Leftrightarrow \varphi_n(t) \to \varphi(t) \Leftrightarrow F_n(x) \to F(x), \qquad (8.11)$$

where φ_n denotes the characteristic function of X_n and F_n the distribution function of X_n.

The Lévy–Lindeberg CLT. If $X_1, \ldots, X_n, \ldots$ is a random sample with $E(X_i) = \mu$ and $\text{Var}(X_i) = \sigma^2$, then

$$S_n^* \equiv \frac{\sqrt{n}(\bar{X}_n - \mu)}{\sigma} = \frac{\sum\limits_{i=1}^{n}(X_i - \mu)}{\sqrt{n}\sigma} \xrightarrow{L} N(0, 1),$$

i.e., for every x,

$$\lim_{n \to \infty} P[S_n^* \leq x] = \Phi(x) = \frac{1}{\sqrt{2\pi}} \int_{-\infty}^{x} e^{-u^2/2} \, du. \qquad (8.12)$$

Lyapunov's CLT. Let $X_1, \ldots, X_n, \ldots$ be a sequence of independent random variables with

$$E(X_i) = \mu_i, \qquad \text{Var}(X_i) = \sigma_i^2, \qquad \beta_i = E|X_i - \mu_i|^3,$$

and set

$$s_n = \left(\sum_{i=1}^{n} \sigma_i^2 \right)^{1/2}, \qquad B_n = \left(\sum_{i=1}^{n} \beta_i \right)^{1/3};$$

if

$$\frac{B_n}{s_n} \to 0 \qquad \text{as} \quad n \to \infty,$$

then

$$S_n^* = \frac{\sum\limits_{i=1}^{n}(X_i - \mu_i)}{s_n} \xrightarrow[n \to \infty]{L} N(0, 1). \qquad (8.13)$$

Necessary and sufficient conditions for the validity of the CLT are given by the

Lindeberg–Feller CLT. Let X_n, μ_n, σ_n, s_n, S_n^* be defined as above, and let F_n be the distribution function of X_n. Then for every $\varepsilon > 0$,

$$\max_{1 \le i \le n} \frac{i}{s_n} \xrightarrow[n \to \infty]{} 0 \quad \text{and} \quad S_n^* \xrightarrow{L} N(0, 1)$$

$$\Leftrightarrow \frac{1}{s_n^2} \sum_{i=1}^{n} \int_{|x-\mu_i| > \varepsilon s_n} (x - \mu_i)^2 \, dF_i(x) \xrightarrow[n \to \infty]{} 0,$$

The classical De Moivre–Laplace CLT for Bernoulli variables (the normal approximation to the binomial distribution) is a special case of (8.12).

Exercises

251. If the independent random variables $X_1, \ldots, X_n, \ldots$ satisfy the condition

$$V(X_i) \le c < \infty, \qquad i = 1, 2, \ldots,$$

then the SLLN holds.

252. In a sequence of random variables $X_1, \ldots, X_n, \ldots$ suppose that X_k depends only on X_{k-1}, X_{k+1}, but that it is independent of all the other random variables ($k = 2, 3, \ldots$). Show that if $V(X_i) \le N < \infty$ ($i = 1, 2, \ldots$), then the WLLN holds.

253.* If for every n, $V(X_i) \le c < \infty$ and $\text{Cov}(X_i, X_j) < 0$ ($i, j = 1, 2, \ldots, n$), then the WLLN holds.

254.* Let $\{X_n\}$ be a sequence of random variables so that $V(X_i) \le c < \infty$ ($i = 1, 2, \ldots$) and $\text{Cov}(X_i, X_j) \to 0$ when $|i - j| \to \infty$ then the WLLN holds (Theorem of Barnstein).

255.* If $X_n \xrightarrow{L} X$, $Y_n \xrightarrow{P} c$, then
(a) $X_n + Y_n \xrightarrow{L} X + c$;
(b) $X_n Y_n \xrightarrow{L} cX$;
(c) $X_n/Y_n \xrightarrow{L} X/c$.

256.* Let $\xi_n = (X_{1n}, X_{2n}, \ldots, X_{kn})'$ ($n = 1, 2, \ldots$) be a sequence of random vectors and $\xi = (X_1, X_2, \ldots, X_k)'$ a random vector with distribution function $F(x_1, \ldots, x_k)$. If

$$\sum_{i=1}^{k} c_i X_{in} \xrightarrow{L} \sum_{i=1}^{k} c_i X_i$$

for all constants $c_1, \ldots, c_k$, then the joint distribution function F_{ξ_n} of $X_{1n}, \ldots, X_{kn}$

has a limit and

$$F_{\xi_n}(x_1, \ldots, x_k) \to F(x_1, \ldots, x_k).$$

257. Let $\{X_n\}$ be a sequence of independent random variables uniformly bounded, that is, there is a constant M such that $P[X_k \le M] = 1$ for all $k = 1$, $2, \ldots$, and suppose that $\mathrm{Var}(X_k) \ne 0$ for every k. If $s_n^2 \to \infty$, then the CLT holds (cf. (8.13)).

258. Show that

$$\lim_{n \to \infty} e^{-n} \sum_{k=0}^{n} \frac{n^k}{k!} = \frac{1}{2}.$$

259. If X and Y are independent Poisson variables with parameters λ_1 and λ_2, respectively, show that

$$\frac{X - Y - (\lambda_1 - \lambda_2)}{(X + Y)^{1/2}} \xrightarrow[\substack{\lambda_1 \to \infty \\ \lambda_2 \to \infty}]{L} N(0, 1).$$

260. Show that for the sequences $\{X_n\}$ of independent random variables with

(a) $P[X_n = \pm 1] = \dfrac{1 - 2^{-n}}{2}$, $P[X_n = \pm 2^n] = \dfrac{1}{2^{n+1}}$, $n = 1, 2, \ldots$,

(b) $P[X_n = \pm n^\lambda] = 1/2$,

the CLT holds.

261. Compare the results given by Chebyshev's inequality and the CLT for the probabilities

$$P[-k < S_n^* \le k] \qquad \text{for } k = 1, 2, 3.$$

262. A variable X has the Pareto distribution with density

$$f(x) = 24x^{-4}, \qquad x \ge 2.$$

Give the graph of the function $g(\delta) = P[|X - \mu| > \delta]$ where $\mu = E(X)$, and compare this with the upper bound given by Chebyshev's inequality.

263. Given that $\mathrm{Var}(X) = 9$, find the number n of observations (the sample size) required in order that with probability less than 5% the mean of the sample differs from the unknown mean μ of X more, (a) than 5% of the standard deviation of X, (b) than 5% of μ given that $\mu > 5$. Compare the answers obtained by using Chebyshev's inequality and the CLT.

264. In a poll designed to estimate the percentage p of men who support a certain bill, how many men should be questioned in order that, with probability at least 95%, the percentage of the sample differs from p less, (i) than 1%, (ii) than 5%, given that, (a) $p < 30\%$, and (b) p is completely unknown.

265. Each of the 300 workers of a factory takes his lunch in one of three competing restaurants. How many seats should each restaurant have so that, on average, at most one in 20 customers will remain unseated?

266. The round-off error to the second decimal place has the uniform distribution on the interval $(-0.05, 0.05)$. What is the probability that the absolute error in the sum of 1,000 numbers is less than 2?

267. The daily income of a card player has uniform distribution in the interval $(-40, 50)$.

(a) What is the probability that he wins more than 500 drachmas in 60 days?

(b) What amount c (gain or loss) is such that at most once in 10 times on average the player will have income less than c during 60 days?

268. In the game of roulette the probability of winning is 18/37. Suppose that in each game a player earns 1 drachma or loses 1 drachma. How many games must be played daily in a casino in order that the casino earns, with probability 1/2, at least 1,000 drachmas daily? Then find the percentage of days on which the casino has a loss.

269. Show that the Γ distribution of Exercise 202 tends to the normal distribution as $s \to \infty$.

CHAPTER 9

Special Topics: Inequalities, Geometrical Probabilities, Difference Equations

Elements of Theory

A. Inequalities

In Chapter 8 several inequalities such as the Chebyshev, Markov, and Kolmogorov inequalities were given. Here we give some inequalities concerning expectations.

1. A function g, defined over the interval (α, β), is called *convex* in (α, β) if for each λ $(0 \le \lambda \le 1)$, and every pair x_1, $x_2 \in (\alpha, \beta)$, we have

$$\lambda g(x_1) + (1 - \lambda)g(x_2) \ge g(\lambda x_1 + (1 - \lambda)x_2). \qquad (9.1)$$

Remark. (a) If g is continuous in (α, β) then the validity of (9.1) for $\lambda = 1/2$ implies the convexity of g.

(b) If the second derivative $g''(x)$ exists, then (9.1) is equivalent to

$$g''(x) \ge 0. \qquad (9.2)$$

The function g is called *concave* if $-g$ is convex.

2. *Jensen's inequality.* Let $g(x)$ be a convex function in (α, β) and $P[X \in (\alpha, \beta)] = 1$. If $E[g(X)] < \infty$, then

$$E[g(X)] \ge g(EX). \qquad (9.3)$$

The proof is based on the observation that there exists a line of support of the curve $y = g(x)$ through the point $(\mu, g(\mu))$ with equation $y = g(\mu) + \lambda(x - \mu)$

such that

$$g(x) \geq g(\mu) + \lambda(x - \mu).$$

3. *Cauchy–Schwarz inequality.* If $E(X^2)$ and $E(Y^2)$ exist then

$$[E(XY)]^2 \leq E(X^2)E(Y^2); \tag{9.4}$$

equality holds only if $Y = \lambda X$ (λ constant). See Exercise 103.

4. *Hölder's inequality.* Let $p > 0$, $q > 0$, such that $p^{-1} + q^{-1} = 1$ and X, Y positive random variables. Then

$$E(XY) \leq [E(X^p)]^{1/p}[E(Y^q)]^{1/q}. \tag{9.5}$$

This gives (9.4) when $p = q = 1/2$.

Proof. The function

$$g(x) = \log x$$

is concave for $x > 0$, i.e., for each λ ($0 \leq \lambda \leq 1$)

$$\lambda \log x_1 + (1 - \lambda) \log x_2 \leq \log(\lambda x_1 + (1 - \lambda)x_2).$$

Hence

$$x_1^\lambda x_2^{1-\lambda} \leq \lambda x_1 + (1 - \lambda)x_2,$$

and setting $\lambda = p^{-1}$, $1 - \lambda = q^{-1}$, $x_1 = X^p/E(X^p)$, $x_2 = Y^q/E(Y^q)$, we obtain

$$\frac{X}{[E(X^p)]^{1/p}} + \frac{Y}{[E(Y^q)]^{1/q}} \leq \lambda \frac{X^p}{E(X^p)} + (1 - \lambda)\frac{Y^q}{E(Y^q)}.$$

Taking expectations gives (9.5).

5. *Minkowski's inequality.* For every integer $k \geq 1$ and any random variables X and Y with $E(X^k) < \infty$, $E(Y^k) < \infty$,

$$[E|X + Y|^k]^{1/k} \leq [E|X|^k]^{1/k} + [E|Y|^k]^{1/k}. \tag{9.6}$$

If we define the norm of X by $\|X\| = [E|X|^k]^{1/k}$, then (9.6) becomes the *triangle inequality*

$$\|X + Y\| \leq \|X\| + \|Y\|.$$

Proof. For $k = 1$, obvious. For $k > 1$, replacing Y by $|X + Y|^{k-1}$ in (9.5), we obtain

$$E|X + Y|^k \leq E(|X||X + Y|^{k-1}) + E(|Y||X + Y|^{k-1})$$
$$\leq [(E|X|^k)^{1/k} + (E|Y|^k)^{1/k}][E|X + Y|^{(k-1)s}]^{1/s}, \tag{9.7}$$

where $k^{-1} + s^{-1} = 1$ and therefore $(k - 1)s = k$. Dividing both sides of (9.7) by $(E|X + Y|)^{1/s}$ we obtain (9.6).

B. Geometrical Probabilities

Geometrical probabilities refer to events corresponding to subsets of a bounded set S of the Euclidean n-space E^n under the assumption that the distribution over S is uniform. Thus the probability of an event

$$A = \{x: x \in S_A \subset S\}$$

is given by

$$P(A) = \frac{\mu(S_A)}{\mu(S)},$$

where $\mu(R)$ denotes the Lebesgue measure (geometric size) of the set R, i.e., its length, area, volume, etc., depending on the dimensionality of R. The solution of problems in geometrical probability contributed a lot to the understanding of the role played by the specification of the random experiment related to a certain event. This is demonstrated, e.g., in the classical *Bertrand paradox* (see Exercise 169).

C. Difference Equations

In many cases a problem in probability can be solved more easily by constructing and solving a related difference equation, i.e., an equation of the form

$$g(x, f(x), \Delta f(x), \ldots, \Delta^r f(x)) = 0,$$

where f is the unknown function and Δ denotes the forward difference operator defined by

$$\Delta f(x) = f(x + h) - f(x) \qquad \text{for some fixed } h > 0,$$

and $\qquad \Delta^r f(x) = \Delta(\Delta^{r-1} f(x)) \qquad \text{for } r \geq 1 \quad (\Delta^0 f \equiv f).$

If $x = 0, 1, 2, \ldots$, then we set

$$y_n = f(n), \qquad n = 0, 1, 2, \ldots,$$

and we have, e.g.,

$$\Delta y_n = y_{n+1} - y_n, \qquad \Delta^2 y_n = y_{n+2} - 2y_{n+1} + y_n.$$

Of special interest in applications are the *linear difference equations with constant coefficients* α_i, i.e., of the form

$$y_{n+r} + \alpha_1 y_{n+r-1} + \cdots + \alpha_{r-1} y_{n+1} + \alpha_r y_n = z_n, \tag{9.8}$$

where n usually ranges over a set of positive integers. If $\alpha_r \neq 0$ the difference equation is said to be of order r. The difference equation is called *non-homogeneous* (complete) or *homogeneous* according to whether $z_n \neq 0$ or $z_n = 0$.

The *characteristic polynomial* $\varphi(\lambda)$ of (9.8) is defined by

$$\varphi(\lambda) = \lambda^r + \alpha_1 \lambda^{r-1} + \cdots + \alpha_{r-1}\lambda + \alpha_r.$$

The solution of the homogeneous equation

$$y_{n+r} + \alpha_1 y_{n+r-1} + \cdots + \alpha_{r-1} y_{n+1} + \alpha_r y_n = 0 \tag{9.9}$$

is based on the

Proposition. (i) With each simple root λ of the characteristic equation $\varphi(\lambda) = 0$ let us associate the special solution $c\lambda^n$ (c any constant) of the homogeneous difference equation (9.9).

(ii) With each real root λ of multiplicity k of $\varphi(\lambda) = 0$ associate the special solution

$$(c_0 + c_1 n + c_2 n^2 + \cdots + c_{k-1} n^{k-1})\lambda^n,$$

where $c_0, c_1, \ldots, c_{k-1}$ are arbitrary constants.

(iii) With each pair of (conjugate) complex roots of length (absolute value) ρ and argument θ associate the special solution

$$\rho^n[A_0 \cos(n\theta + B_0) + \cdots + A_{k-1} n^{k-1}(\cos(n\theta + B_{k-1}))],$$

where k is the multiplicity of the complex root and $A_0, \ldots, A_{k-1}, B_0, \ldots, B_{k-1}$ arbitrary constants.

The sum of the above special solutions of (9.9) determine the general solution of (9.9) containing r arbitrary constants which are determined by the *initial* or *boundary conditions* satisfied by (9.9).

General solution of the complete equation (9.8). This can be found by adding a special solution of (9.8) to the general solution of the homogeneous (9.9).

Example. Solve the nonhomogeneous equation of the second order

$$y_{n+2} - y_{n+1} - 6y_n = 2^n, \qquad n = 0, 1, \ldots. \tag{9.10}$$

The roots of the characteristic equation

$$\varphi(\lambda) = \lambda^2 - \lambda - 6 = 0$$

are $\lambda_1 = 3$, $\lambda_2 = -2$, and by the preceding proposition the general solution of the homogeneous equation

$$y_{n+2} - y_{n+1} - 6y_n = 0,$$

is

$$y_n^* = c_1 3^n + c_2(-2)^n,$$

with c_1, c_2 arbitrary constants.

For the general solution of (9.10) we must first find a special solution y_n^0

which is to be added to y_n^*. We consider the special solution

$$y_n^0 = c2^n, \tag{9.11}$$

where c has to be determined to satisfy (9.10); substituting (9.11) into (9.10) gives

$$c = -\tfrac{1}{4}, \quad \text{i.e.,} \quad y_n = -(\tfrac{1}{4})2^n.$$

Thus the general solution of (9.10) is

$$y_n = y_n^* + y_n^0 = c_1 3^n + c_2(-2)^n - (\tfrac{1}{4})2^n.$$

Exercises

A. Inequalities

270. If $g(x) \geq 0$ for every x and $g(x) \geq c$ for $x \in (\alpha, \beta)$, then

$$P[X \in (\alpha, \beta)] \leq c^{-1}E[g(x)].$$

271. (*Continuation*). Show that for every constant $t > 0$

$$P[X > t] \leq \frac{1}{(t + c)^2} E(X + c)^2.$$

272. If X_1 and X_2 are independent and identically distributed random variables then for every $t > 0$

$$P[|X_1 - X_2| > t] \leq 2P[|X_1| > \tfrac{1}{2}t].$$

273. *Cantelli's inequality.* One tail Chebyshev's inequality. If $E(X) = \mu$, $\mathrm{Var}(X) = \sigma^2$, then

$$P[X - \mu \leq c] \leq \frac{\sigma^2}{\sigma^2 + c^2}, \quad c < 0,$$

$$\geq \frac{c^2}{\sigma^2 + c^2}, \quad c \geq 0,$$

(see Exercise 271).

274.* *Gauss inequality.* If x_0 is the mode of a continuous unimodal distribution and $\tau^2 = E(X - x_0)^2$ then

$$P[|X - x_0| \geq \lambda\tau] \leq \frac{4}{9\lambda^2}.$$

275. If $g(x) \geq 0$ and even, i.e., $g(x) = g(-x)$ and in addition $g(x)$ is non-decreasing for $x > 0$, show that for every $c > 0$

$$P[|X| \geq c] \leq \frac{Eg(x)}{g(c)}.$$

276. (*Continuation*). If $g(x)$ of Exercise 275 satisfies $|g(x)| \le M < \infty$, then

$$P[|X| \ge c] \ge \frac{Eg(X) - g(c)}{M}.$$

277. (*Continuation*). Let g be as in Exercise 275 and $P[|X| \le M] = 1$ then

$$P[|X| \ge c] \ge \frac{Eg(X) - g(c)}{g(M)}.$$

278.* *Berge's inequality.* Let $E(X_i) = \mu_i$, $\text{Var}(X_i) = \sigma_i^2$ $(i = 1, 2)$ and $\rho(x_1, x_2) = \rho$, then

$$P\left[\max\left(\frac{|X_1 - \mu_1|}{\sigma_1}, \frac{|X_2 - \mu_2|}{\sigma_2} \right) > c \right] \le \frac{1 + \sqrt{1 - \rho^2}}{c^2}.$$

279.* *Jensen's inequality for symmetric matrix functions* (Cacoullos and Olkin, *Biometrika*, 1965). The symmetric matrix function $G(X)$ of the matrix X of order $m \times n$ is called convex if, for every pair X_1, X_2 of $m \times n$ matrices and $0 \le \lambda \le 1$, by analogy to (9.1), it satisfies

$$G(\lambda X_1 + (1 - \lambda)X_2) \le \lambda G(X_1) + (1 - \lambda)G(X_2),$$

where for symmetric matrices X, Y we write $X \ge Y$ if $X - Y$ is a nonnegative definite matrix. Let X be a random matrix, i.e., a matrix whose elements are random variables and G a convex matrix function. Then $G(EX) \le EG(X)$.

280. For a positive random matrix show that

$$E(X^{-1}) \ge (E(X))^{-1}.$$

281. Show that if $E(Y) < \infty$ and $Y > 0$ then

$$E \log(Y) \le \log E(Y).$$

282. For independent and identically distributed random variables $X_1, X_2,$ $\ldots, X_n$ with $P[X_i > 0] = 1$ and $\text{Var}(\log X_i) = \sigma^2$ show that for every $\varepsilon > 0$

$$P[e^{n[E(\log Y_i) - \varepsilon]} < X_1 X_2 \ldots X_n < e^{n[E(\log Y_i) + \varepsilon]}] \ge 1 - \frac{\sigma^2}{n\varepsilon^2}.$$

Hence deduce

$$P[X_1 X_2 \ldots X_n < (EY_i)^n e^{n\varepsilon}] \ge 1 - \frac{\sigma^2}{n\varepsilon^2}.$$

B. Geometrical Probabilities

283. A point is chosen at random on each of two consecutive sides of a rectangle. What is the probability that the area of the triangle bounded by the two sides of the rectangle and the line segment connecting the two points is less than 3/8 of the area of the rectangle?

284. Two points are selected at random on segment AB which is thus divided into three parts. What is the probability that the three line segments can form a triangle?

285. What is the probability that three points selected at random on the circumference of a circle lie on a semicircle?

286. A line segment of length 1 is divided into five parts by four points selected at random on it. What is the probability that each part is less than $1/2$?

287.* A line segment AB is divided by a point C into two parts $AC = a$ and $CB = b$. Two points X and Y are taken on AC and CB, respectively. What is the probability that AX, XY, BY can form a triangle?

288.* *Buffon's needle problem.* A smooth table is given with equidistant parallel lines at distance $2a$. A needle of length 2μ ($\mu < a$) is dropped onto the table. What is the probability that the needle crosses a line?

289.* A closed convex curve with diameter less than $2a$ is dropped onto the table of Exercise 288. Find the probability that a line is crossed by the curve.

C. Difference Equations

290. A has a 10-drachma coins and B has b 5-drachma coins. Each of them takes a coin out of his pocket and gives it to the other. This is repeated n times. What is the probability that A gives a 10-drachma coin to B on the nth exchange?

291. A tosses a die until an ace or a six appears twice.
(a) What is the probability that the game terminates with the appearance of an ace after no more than n tosses?
(b) What is the probability that the game ends with the appearance of an ace?

292. A professor has three lecture rooms A, B, and C available each year for teaching a course. Since he wishes to use a different classroom every year, at the end of the year he chooses one of the other two rooms for the next academic year as follows. The first year he teaches in A and selects between B and C by tossing a coin. If heads appears he chooses B, otherwise he selects C. At the end of each year he makes his selection in the same way. Find the probability p_n that he teaches in A during the nth year. Similarly for B and C.

293. Find the probability p_n that in a sequence of n Bernoulli trials, with probability of success (S) p the pattern SF does not appear at all ($F \equiv$ failure).

294. Let y_k be the expected number of occupied cells when placing (at random) k balls into n cells. Find y_k:
(a) by using difference equations;
(b) by referring to the expected value of a binomial.

295. A pair of dice is thrown n times; at each throw we observe whether a double six appears or not. Show that the probability p_n of an even number of double six's satisfies the difference equation

$$p_n - \frac{17}{18}p_{n-1} = \frac{1}{36}.$$

Hence find p_n.

296.* Each of n urns contains w white balls and b black balls. One ball is transferred from the first urn to the second urn, then another ball from the second urn is transferred to the third one, and so on. Finally, we draw a ball from the nth urn. What is the probability that it will be white?

297. A and B play the following game. A throws a coin I with probability of heads p until he gets tails; then B begins throwing another coin II with probability of heads p' until he gets tails. Then A begins again and so on. Given that A begins the game, what is the probability of heads at the nth throw?

298. A throws two dice and he gets r points if r aces appear ($r = 0, 1, 2$); the game stops when no ace appears for the first time. Find the probability p_n that he has n points at stopping. If the game continues indefinitely, what is the probability that A has n points at some point of the game?

299. A box contains a good lamps and b bad lamps. At each draw of a lamp, if it is good it is kept in the box, otherwise it is replaced by a good one (from another box). What is the expected number of good lamps in the box after n trials? Hence calculate the probability of selecting a good lamp at the nth trial.

300. Let p_k denote the probability that no cell takes more than two balls when placing k balls in n cells at random. Find, (a) p_k (b) $\lim_{k \to \infty} p_k$ and $n = $ constant, (c) $\lim p_k$ for $k = $ constant and $n \to \infty$.

CHAPTER 10

General Exercises

301.* Using the Cauchy–Schwarz inequality show that
$$g(t) = \log E(|X|^t)$$
is a convex function. Then show that $[E(|X|^t)]^{1/t}$ is an increasing function of t.

302.* Every real characteristic function $\varphi_0(t)$ satisfies the inequality
$$1 - \varphi_0(2t) \le 4(1 - \varphi_0(t)).$$

303.* Every real characteristic function $\varphi_0(t)$ satisfies
$$1 + \varphi_0(2t) \ge 2\{\varphi_0(t)\}^2.$$

304. If X_1, X_2, X_3, X_4 are jointly normal with $\mathrm{Cov}(X_i, X_j) = E(X_i X_j) = \sigma_{ij}$, show by the use of moment generating functions, that
$$E(X_1 X_2 X_3 X_4) = \sigma_{12}\sigma_{34} + \sigma_{14}\sigma_{23} + \sigma_{13}\sigma_{24}.$$

305. Let $X_1, X_2, \ldots, X_n$ be a random sample from the uniform distribution in $(0, 1)$ and let $Y_n = n \min(X_1, X_2, \ldots, X_n)$. Using characteristic functions, show that the asymptotic distribution $(n \to \infty)$ of Y_n is the exponential e^{-y}, $y > 0$.

306. Let $(X_1, Y_1), \ldots, (X_n, Y_n)$ be a random sample from the bivariate normal with means 0, variances 1, and correlation ρ. Let the random variable
$$Z_j = \begin{cases} 1 & \text{if } (X_j - \bar{X})(Y_j - \bar{Y}) > 0, \\ 0 & \text{if } (X_j - \bar{X})(Y_j - \bar{Y}) < 0, \end{cases}$$
Find $E(\bar{Z})$.

307.* n tickets are numbered 1 through n. Of these, r tickets drawn at random are the lucky ones. A lucky ticket receives an amount equal to the number it bears. What is the expected amount given on every draw? What is the variance of this amount?

308. In the generalized binomial with $P[X_i = 1] = p_i$ $(i = 1, 2, \ldots, n)$, show that as $n \to \infty$ the distribution of the number of successes $X = \sum_{i=1}^{n} X_i$ tends to the Poisson with parameter λ where $\lambda = \lim_{n \to \infty} \sum_{i=1}^{n} p_i$ assuming $\max_i \lambda_i \to 0$.

309. Let X be the number of failures preceding the nth success in a sequence of independent Bernoulli trials with probability of success p. Show that if as $n \to \infty$, $q = 1 - p \to 0$, so that $nq \to \lambda$, then

$$P[X = k] \to e^{-\lambda} \frac{\lambda^k}{k!}.$$

310. If the continuous random variable X is such that for some integer $k > 0$, kX has the Gamma distribution and the conditional distribution of Y given X is Poisson, then Y has a negative binomial distribution.

311. Let $X_1, X_2, \ldots, X_n$ be a random sample from the Laplace distribution

$$f(x) = \frac{1}{2\sigma} e^{-(|x-\mu|)/\sigma}, \qquad -\infty < x < \infty.$$

By using characteristic functions show that the random variable

$$Y = \frac{1}{n} \sum_{j=1}^{n} |X_j - \mu|$$

is distributed as $(\sigma/2n)\chi^2_{2n}$. Hence deduce that $E(Y) = \sigma$.

312. Let $X_1, X_2, \ldots, X_n$ be a random sample from a distribution with $E(X_i) = 0$ and $\mathrm{Var}(X_i) = 1$. Show that

$$Y_n = \frac{\sqrt{n}S_n}{\sum\limits_{i=1}^{n} X_i^2} \xrightarrow[n \to \infty]{} N(0, 1),$$

$$Z_n = \frac{S_n}{\sqrt{\sum\limits_{i=1}^{n} X_i^2}} \xrightarrow[n \to \infty]{} N(0, 1),$$

where $S_n = X_1 + X_2 + \cdots + X_n$.

313. Let $X_{(1)}, \ldots, X_{(n)}$ be the ordered sample from a uniform distribution in $(0, 1)$ and $Y_r = X_{(1)} + X_{(2)} + \cdots + X_{(r)}$. Show that the conditional distribution of $W = Y_r/Y_{(r+1)}$ is the distribution of the sum of r observations of a random sample from the same uniform distribution.

314.* N points $X_1, X_2, \ldots, X_n$ are selected at random in $(0, 1)$. Show that the distribution of the sum $S_n = X_1 + X_2 + \cdots + X_n$ has density

$$f_n(x) = \frac{1}{(n-1)!}\left\{x^{n-1} - \binom{n}{1}(x-1)^{n-1} + \cdots\right\}$$

$$= \frac{n}{(n-1)!}\left\{\frac{(n-x)^{n-1}}{n} - \binom{n-1}{1}\frac{(n-1-x)^{n-1}}{n-1} + \cdots\right\},$$

where the sum extends as long as $x - 1, x - 2, \ldots$ and $(n - 1 - x), (n - 2 - x),$ $\ldots$ are positive.

315. Let $X_1, X_2, \ldots, X_n$ be independent discrete variables with the same uniform distribution on the set of integers $0, 1, \ldots, m - 1$, i.e.,

$$P[X_1 = j] = \frac{1}{m} \qquad \text{for} \quad j = 0, \ldots, m - 1.$$

Find the probability generating function of the sum $S_n = X_1 + \cdots + X_n$ and hence the probability function of S_n (Feller, 1957, Exercise 18, page 266).

316. *(Continuation).* A die is thrown three times. What is the probability that the sum of the outcomes is 10?

317.* *Generalized Banach match box problem.* A chain smoker has $m + 1$ match boxes; each time he lights a cigarette he chooses one of the boxes at random. Suppose that (to begin with) every box contains N matches. Let X_1, $X_2, \ldots, X_m$ be the numbers of matches left in the remaining m boxes, when:
(a) a box is found empty for the first time;
(b) any box is emptied first.
Note: The case $m = 1$ is known as the Banach match box problem (Feller, *ibid*, p. 157).

318. *(Continuation).* Suppose that box C_k $(k = 1, \ldots, m)$ is chosen with probability p_k. What is the probability that C_k empties first?

319. Show that X_1, X_2, X_3 with a joint discrete distribution are independent if and only if

$$P[X_3 = x_3 | X_2 = x_2, X_1 = x_1] = P[X_3 = x_3]$$

for all (x_1, x_2) with $P[X_1 = x_1, X_2 = x_2] > 0$ and

$$P[X_2 = x_2 | X_1 = x_1] = P[X_2 = x_2] \qquad \text{for all } x_1 \text{ with } P[X_1 = x_1] > 0.$$

320. Let $X_1, X_2, \ldots, X_n$ be independent and identically distributed positive random variables with $E(X_i) = \mu$, $E(X_i^{-1}) = r$. Let $S_n = X_1 + X_2 + \cdots + X_n$. Show that $E(S_n^{-1})$ exists and

$$E\left(\frac{X_i}{S_n}\right) = \frac{1}{n}, \qquad i = 1, \ldots, n.$$

(Feller, *ibid*, Exercise 36, page 226.)

321. (*Continuation*). Show that

$$E\left(\frac{S_m}{S_n}\right) = \frac{m}{n} \quad \text{for } m \le n, \qquad E\left(\frac{S_m}{S_n}\right) = 1 + (m - n)\mu E(S_n^{-1}) \quad \text{for } m > n.$$

322. A caterpillar is moving on the edges of a tetrahedron $ABCD$ on whose top there is glue. In a unit of time the caterpillar goes from any vertex (except D) to any other vertex with the same probability $1/3$. Suppose that the caterpillar at time $t = 0$ is on the vertex A. What is the probability that
(a) the caterpillar finally gets stuck;
(b) the caterpillar finally gets stuck coming from vertex B?

323. A flea moves at random on a plane with leaps of constant length a and in a random direction at each step. If it starts from the origin, let (X_n, Y_n) be the position of the flea after n jumps. Where

$$X_n = \sum_{i=1}^{n} a \cos \theta_i, \qquad Y_n = \sum_{i=1}^{n} a \sin \theta_i.$$

The angles θ_i are independently uniform in $(0, 2\pi)$.
(a) Find, (i) $E(X_n)$, $E(Y_n)$, $\text{Var}(X_n)$, $\text{Var}(Y_n)$, (ii) $E(R_n^2)$ where $R_n^2 = X_n^2 + Y_n^2$.
(b) Show that the X_n, Y_n are uncorrelated but not independent.
(c) Find the distribution of (X_n, Y_n) for large n ($n \to \infty$) and the density of R_n. What is the expected distance of the flea from the origin for large n?

324.* *A nonnormal distribution with all marginals normal.* Let the random variables $X_1, \ldots, X_n$ have the joint density

$$f(x_1, \ldots, x_n) = \frac{1}{(2n)^{n/2}} \exp\left(-\frac{1}{2}\sum_{i=1}^{n} x_i^2\right)\left[1 + \prod_{i=1}^{n} (x_i \exp(-\tfrac{1}{2}x_i^2))\right].$$

Show that any $n - 1$ of the $X_1, \ldots, X_n$ are independent normal with mean 0 and variance 1. Are the $X_1, \ldots, X_n$ jointly normal?

325.* In Exercise 288 suppose the plane is divided into squares by equidistant lines parallel to the axes. Let $2a$ be the side of each square and 2μ the length of the needle. Find the average number of lines crossed by the needle, (a) $\mu > a$, (b) when μ is arbitrary.

326.* Let X, Y be independent χ^2 variables each with n degrees of freedom. Show that

$$Z = \frac{\sqrt{n}}{2} \frac{X - Y}{\sqrt{XY}}$$

has the Student distribution with n degrees of freedom. Hence deduce that:

(a) $$t_n = \frac{\sqrt{n}}{2}\left(\sqrt{F_{n,n}} - \frac{1}{\sqrt{F_{n,n}}}\right),$$

where $F_{m,n}$ denotes an F variable with m and n degrees of freedom.

(b) If $t_n(a)$ and $F_{n,n}(a)$ are the upper percentile points of t_n and $F_{n,n}$, respectively, express one in terms of the other.

327. If for a given value λ of Λ the random variable $X(\Lambda)$ is Poisson with parameter λ and Λ is a random variable with a gamma density

$$\Gamma(\lambda) = \frac{a^\rho}{\Gamma(\rho)} \lambda^{\rho-1} e^{-a\lambda},$$

show that $X(\Lambda)$ has the negative binomial distribution

$$p_k = P[X(\Lambda) = k] = \binom{-\rho}{k}(-p)^k q^\rho, \qquad k = 0, 1, \ldots,$$

where $p = (1 + a)^{-1}$, $q = a(a + 1)^{-1}$.

328.* Show that for any two random variable(s) X and Y with $\mathrm{Var}(X) < \infty$,

$$\mathrm{Var}(X) = E[\mathrm{Var}(X|Y)] + \mathrm{Var}[E(X|Y)].$$

329.* Using the distribution of $X_{(t)}$ where $X_{(1)}, \ldots, X_{(n)}$ is the ordered sample from the uniform in $(0, 1)$, show the following relation between the incomplete B function $I_x(n_1, n_2)$ and the binomial distribution function:

$$\sum_{k=r}^{n} \binom{n}{k} p^k q^{n-k} = I_p(r, n - r + 1)$$

$$= \frac{1}{B(r, n - r + 1)} \int_0^p x^{r-1}(1 - x)^{n-r} \, dx.$$

SUPPLEMENTS

Miscellaneous Exercises

I-1. For any two events A, B show that the following relationships are equivalent: $A \subset B$, $A' \supset B'$, $A \cup B = B$, and $A \cap B' = \emptyset$.

I-2. Show that $(A \triangle B)' = (AB) \cup (A'B')$ and $P(A \triangle B) = P(A) + P(B) - 2P(AB)$, where $\triangle$ is the symmetric difference, defined in Exercise 34.

I-3. Let n be the number of elements in the sample space Ω and $N(E)$ the number of elements in the event E. Show that the data: $N = 1{,}000$, $N(A) = 525$, $N(B) = 312$, $N(C) = 470$, $N(AB) = 42$, $N(AC) = 147$, $N(BC) = 86$, and $N(ABC) = 25$ are inconsistent.
   Hint: Check $N(A \cup B \cup C)$.

I-4. (Lewis Caroll, *A Tangled Tale*, 1881). In a fierce battle, at least 70% of the soldiers lost an eye, at least 74% lost an ear, at least 80% lost an arm, and at least 85% lost a leg. What is the smallest possible percentage of soldiers that each lost an eye, an ear, an arm, and a leg?
   Answer: 10%.

I-5. In a market research survey, from a sample of 1,000 smokers, 811 said that they prefer brand "A", 752 prefer "B", and 418 prefer "C". Furthermore, 570 said that they like both "A" and "B", 356 that they like both "A" and "C", 348 that they like both "B" and "C", and 297 that they like all three brands. How did the researcher discover that the results were inconsistent?
   Hint: How many prefer at least one of the three brands?

I-6. We throw two dice once. Let A: doubles, B: at least one die shows 5 or 6, C: the sum of the outcomes of the two dice is less than 6.
   (a) Find $P(A)$, $P(B)$, $P(C)$, $P(AB)$, and $P(AC)$.
   (b) Using only the results of (a) calculate: $P(A \cup B)$, $P(B \cup C)$, and $P(A \cup C)$.
   Answer: (a) $1/6$, $5/9$, $5/18$, $1/18$, $1/18$. (b) $2/3$, $5/6$, $7/18$.

I-7. Three athletes are participating in the Marathon. Their chances of finishing under 2 hours are 1/3, 1/5, and 1/12, respectively. Their times for the duration of the course are independent. What is the probability that at least one finishes in less than 2 hours?
Answer: 23/45.

I-8. A horse has probability p of jumping over a specific obstacle. Given that in 5 trials it succeeded 3 times, what is the conditional probability that it had succeeded in the first trial?
Answer: 3/5.

I-9. In a train car 3 seats are facing the front and 3 are facing the rear. Two women and three men enter the car and sit at random. What is the probability:
(a) that the two women are facing each other?
(b) that two men are facing each other?
Answer: (a) 1/10. (b) 3/10.

I-10. On the basis of the following data, examine the relationship between the father having brown eyes (event A) and the son having brown eyes (event B). The percentages observed were: AB: 5%, AB': 7.9%, $A'B$: 8.9%, and $A'B'$: 78.2%.
Answer: $P(B/A) = 0.39$ and $P(B/A') = 0.10$.

I-11. We throw three dice. What is the probability that at least one 6 appears given that the three outcomes were completely different. Generalize the result for the case of a regular polyhedral die.
Answer: $\binom{5}{2} \Big/ \binom{6}{3}$.

I-12. Ninety percent of a population have a television set (event A), while 80% have a car (event B). What is the minimum percentage of the population with a car that also have a television?
Answer: $P(A/B) \geq 87.5\%$.

I-13. Let $P(A) = \alpha$ and $P(B) = 1 - \varepsilon$ where ε is a small number. Give the upper and lower bound of $P(A/B)$.
Answer: $(\alpha - \varepsilon)/(1 - \varepsilon) \leq P(A/B) \leq (1 - \varepsilon)$.

I-14. Let X_1, X_2 be independent random Bernoulli variables with the same parameter p and let $Y = 0$ if $X_1 + X_2 =$ even and $Y = 1$ if $X_1 + X_2 =$ odd.
(a) For what values of p are X_1 and Y independent?
(b) Using (a), define three events A, B, and C such that they are pairwise independent but not completely independent.
Answer: (a) $p = 1/2$. (b) $A = \{X_1 = 0\}$, $B = \{X_2 = 0\}$, $C = \{Y = 0\}$.

I-15. We throw three dice twice. What is the probability of having the same result the second time when the dice are, (a) not distinguishable? (b) distinguishable?
Answer: (a) 1/56. (b) $(6 + 90 \times 3 + 120 \times 6)/6^6$.

I-16. At a party, A is introduced to 6 women and 4 men. What is the probability that he identifies all the couples correctly if he knows that there are, (a) 4 couples, (b) 3 couples?

Answer: (a) 1/360. (b) 1/480.

I-17. A young person driving a motorcycle on a highway with 4 lanes changes lane every 15 seconds. What is the probability p of finding himself in the lane in which he started 1 minute later?

Answer: The answer depends on whether the initial lane is an interior or an exterior one. In the first case $p = 5/8$, and in the second case $p = 1/5$.

I-18. In city A it rains half of the days of the year and the weather forcast is correct with probability 2/3. Mr. Sugar worrying about getting wet always takes his umbrella when the forcast calls for rain and with probability 1/3 otherwise. Find the probability that:

(a) he is caught in the rain without his umbrella;

(b) he has his umbrella with him when there is no rain.

Answer: (a) 2/9. (b) 5/9.

I-19. The telegram signals ($\cdot$) and ($-$) are sent with proportion 3:4. Due to faults in transmission a dot ($\cdot$) becomes ($-$) with probability 1/3. If a signal has arrived as ($\cdot$) what is the probability that it was transmitted correctly?

Answer: 43/84 (Bayes formula).

I-20. A die is thrown 10 times and let the maximum outcome be M and the minimum m. Find the $P(m = 2, M = 5)$.

Hint: Consider the probability $P(m \geq 2, M \leq 5)$.

Answer: $(4/6)^{10} - 2(3/6)^{10} + (2/6)^{10}$.

I-21.* *A theorem of Poisson.* An urn contains b black balls and w white balls. n balls are drawn from the urn at random, their color unnoticed, then m additional balls are drawn. Show that the probability that there are k black balls among the m balls is the same as if the m balls were chosen from the beginning (i.e., before choosing the n balls).

In other words, *in sampling from a finite population the final (hypergeometric) probabilities are not affected by the content of an initial sample when we do not know that content.* Obviously, if we know the composition of the first sample (i.e., the colors of the n balls), then, in general, the probabilities are affected.

Hint: Verify first the case $m = 1$, calculating the probability $P(A_n)$ of choosing a black ball on the $n + 1$ trial; show that $P(A_n) = b/(b + w)$ with $0 \leq n \leq b + w - 1$.

I-22. After an earthquake the books of a large library fall on the floor and the servant, who does not know how to read, reshelves them in any order.

(a) What is the probability that exactly k books are placed back in their initial position?

(b) How many books are expected to be reshelved in their initial position?
Answer: (a) $e^{-1}/k!$. (b) 1.

I-23. *John Smith's problem*. John Smith in 1693 posed the following problem: Is the probability for each of the three players to win the same if: the first must obtain at least one ace throwing 6 dice, the second at least two aces throwing 12 dice, and the third at least three aces throwing 18 dice? Newton and Tollet found that the first player has a better chance than the second one, and the second player has a better chance than the third one.

I-24. A homogeneous and regular die with v faces is thrown n times. Let $p(n, k)$ be the probability that a given face appears less than k times. Show that:
(a) $p(vn, n)$ is a decreasing function of v for given n;
(b) $p(vn, n) < 1/2$;
(c) $p(2n, n) \to 1/2$ as $n \to \infty$ (De Moivre central limit theorem).

I-25. The Massachusetts State Lottery issues 1 million tickets. The first winning ticket wins $50,000, the next 9 winning tickets win $2,500 each, the next 90 tickets win $250 each, and 900 tickets win $25 each. What is the expected gain of a person buying, (a) one ticket? (b) ten tickets?
Answer: $E(\sum_{i=1}^{n} X_i) = \sum_{i=1}^{n} E(X_i)$. (a) $0.1175. (b) $1.175.

I-26. *(Continuation)*. What is the answer to (a) and (b) if:
(i) the state issues 2 million tickets?
(ii) the winning tickets win twice as much?

I-27. In a survey on the housing problem, 2 apartments are chosen from each of 4 buildings (having both owner-occupied as well as rented apartments). If we know that:

Building No.	1	2	3	4
Owner-occupied	5	4	8	2
Rented	8	10	9	10

What is the expected number of owner-occupied apartments in the above sample?
Answer: $2(5/13 + 4/14 + 8/17 + 2/10)$. Use additivity of expectations.

I-28. Out of 365 people:
(a) What is the probability that at least two have their birthday on 2 April?
(b) What is the expected number of people having their birthday on 2 April?
(c) What is the expected number of days in a year that are birthdays: (i) for at least one person? (ii) for at least two persons?
How can you approximate the answers to the above questions?

Answer:
(a) $1 - (364/365)^{365} - 365(364)^{364}/365^{365} \simeq 1 - 2e^{-1} = 0.26$.
(b) 1.

(c) (i) $365\left\{1 - \left(\dfrac{364}{365}\right)^{365}\right\} \simeq 365(1 - e^{-1}) = 231$.

(ii) $365p$ where p was given in (a).

I-29. The duration T of a telephone call satisfies the following equation:

$$P[T > t] = p \cdot \exp[-\lambda t] + (1 - p) \cdot \exp[-\mu t] \qquad \text{for} \quad t > 0,$$

where $0 \le p \le 1$, $\lambda > 0$, $\mu > 0$. Find $E(T)$ and $\text{Var}(T)$.
Answer: $E(T) = p/\lambda + (1 - p)/\mu$. $\text{Var}(T) = 2p/\lambda^2 + 2(1 - p)/\mu^2 - [E(T)]^2$.

I-30.* *The generalized problem of points by Pascal.* Two gamblers A and B play a series of independent games in each of which A has probability p of winning and B has probability $q = 1 - p$ of winning. Suppose that at a given moment the series of games is interrupted and A needs m additional points and B needs n additional points to be the winner. The bet should be split between A and B in the ratio $P(A):P(B)$, where $P(A) = 1 - P(B)$ is the probability that A eventually wins. Show that

$$P_A = \sum_{k=m}^{m+n-1} \binom{m+n-1}{k} p^k - q^{m+n-1-k} = \sum_{m=0}^{n-1} \binom{m+k-1}{k} p^m q^k.$$

Note: The above solution was first obtained by Montmort (1675–1719). Explain the idea behind each term in the expressions for P_A. Try to show the equality between those two expressions for P_A using the properties of binomial coefficients (see Chapter 1).

I-31. An employee leaves between 7:30 a.m. and 8:00 a.m. to go to his office, and he needs 20–30 minutes to get to the railway station. The departure time and the duration of travel are distributed independently and uniformly in the corresponding intervals. He can take either of the following two trains: the first leaves at 8:05 a.m. and takes 35 minutes; and the second leaves at 8:25 a.m. and takes 30 minutes.
(a) What time, on average, does he arrive at his office?
(b) What is the probability that he misses both trains?
(c) How many times is he expected to miss both trains in 240 working days?
Answer: (a) 8:50. (b) 1/24. (c) 10.

I-32. Show that if X and Y are independent, then (cf. Exercise 328)

$$\text{Var}(XY) = \text{Var}(X) \cdot \text{Var}(Y) + \text{Var}(X) \cdot [E(Y)]^2 + \text{Var}(Y) \cdot [E(X)]^2.$$

Conclude that if $X_1, \ldots, X_n$ are completely independent with means zero, then

$$\text{Var}\left(\prod_{i=1}^{n} X_i\right) = \prod_{i=1}^{n} \text{Var}(X_i).$$

I-33. We toss a fair coin until two consecutive heads or tails appear. What is the probability that an even number of tosses will be required?
Answer: 2/3.

I-34.* *Maxwell–Boltzmann statistics.* s distinguishable balls are placed randomly into n different cells. Find the probability that exactly r cells will be occupied.
Answer:

$$n^{-s}\binom{n}{r}\left[r^s - \binom{r}{1}(r-1)^s + \binom{r}{2}(r-2)^2 - \cdots + (-1)^{r-1}\binom{r}{r-1}1^s\right].$$

Hint: The probability that exactly k events occur among n events $A_1, \ldots, A_n$ is given by (see, e.g., Feller, 1957),

$$S_r - \binom{r+1}{1}S_{r+1} + \cdots + (-1)^r\binom{n}{r}S_n$$

(cf. (1.1)).

I-35. A large board is divided into equal squares of side a. A coin of diameter $2r < a$ is thrown randomly onto the board. Find the probability that the coin overlaps, (i) with exactly one square, (ii) with exactly two squares.
Answer: (i) $(1 - 2r/a)^2$. (ii) $1 - 4(r/a)^2$.

I-36. A square of side a is inscribed into a circle. Find the probability that:
(i) any randomly selected point of the circle will be an interior point of the square.
(ii) of eight randomly selected points in the circle, three fall inside the square, two in one of the four remaining parts of the circle, and one in each of the other three parts of the circle.
Answer: (i) $2/\pi$. (ii) $(8!\,4/3!\,2!\,1!\,1!\,1!)(2/\pi)^3(1 - 2/\pi)^5(1/4)^5$.

I-37. The distance between two successive cars on a highway follows the exponential distribution with mean $20m$. Find the probability that there are 90 to 110 cars on a 2-km stretch of the highway.
Answer: $e^{-100}\sum_{k=90}^{100}100^k/k! \approx 0.68$.

I-38. The number of cars crossing a toll bridge follows the Poisson distribution with an average of 30 cars per minute. Find the probability that in 200 seconds no more than 100 cars will pass through.
Answer: The sum of independent exponentials is Erlang (i.e., gamma with $s = 100$, see (2.11));

$$\frac{1}{99!\,2^{100}}\int_{200}^{\infty}t^{99}e^{-t/2}\,dt = \sum_{k=0}^{99}e^{-100}\frac{100^k}{k!}.$$

I-39. (*Continuation*). Establish the relation

$$\frac{\lambda^n}{(n-1)!}\int_t^{\infty}e^{-\lambda x}x^{n-1}\,dx = \sum_{k=0}^{n-1}e^{-\lambda t}\frac{(\lambda t)^k}{k!}, \tag{*}$$

between the Erlang distribution function and the Poisson distribution function.

Hint: The probability that less than n events occur in time t, according to a Poisson distribution (process) $X(t)$ with

$$P[X(t) = k] = e^{-\lambda t} \frac{(\lambda t)^k}{k!},$$

is equal to the probability that the waiting time W_n until the nth event exceeds t. Hence, also conclude that the density of W_n is the Erlang density on the left side of $(*)$ above.

I-40. (*Continuation*). Show that

$$\frac{1}{(n-1)!} \int_\lambda^\infty e^{-x} x^{n-1} \, dx = \sum_{k=0}^{n-1} e^{-\lambda} \frac{\lambda^k}{k!}.$$

I-41. Define $\Phi(a; b) = \Phi(b) - \Phi(a)$ for $0 < a < b$ with Φ the distribution function of $N(0, 1)$. Show that $\Phi(0; 1) > \Phi(1; 2)$. Generalize this result for any equal intervals.
Answer: $\Phi(a + x) - \Phi(a) > \Phi(b + x) - \Phi(b); x > 0$.

I-42. Using the fact that if X follows the normal distribution $N(0, 1)$, then

$$P[X^2 \le x] = (2/\pi)^{1/2} \int_0^{\sqrt{x}} e^{-u^2/2} \, du,$$

conclude that

$$\Gamma(1/2) = \int_0^\infty x^{(1/2)-1} e^{-x} \, dx = \sqrt{\pi}.$$

I-43. The probability of hitting a target with one shot is p_1 and the probability of destroying the target with $k(k \ge 1)$ shots is $1 - p_2^k$. Find the probability of destroying the target after n shots.
Answer: $1 - [1 - p_1(1 - p_2)]^n$.

I-44. Let X follow a Poisson distribution with parameter λ. We select N random points from the interval $[0, 1]$. Let X_i be the number of points in the interval $[(i - 1)/n, i/n]$ $(i = 1, \ldots, n)$. Show that the random variables X_i are independent and identically distributed.

I-45. If X and Y are independent with $P[X = 0] = P[X = 1] = 1/2$ and $P[Y \le y] = y$ $(0 \le y \le 1)$, find the distribution of the following random variables W, Z where:
(i) $W = X + Y$;
(ii) $Z = XY$.
Answer:
(i) $f(w) = 0.5, \quad 0 < w < 2$;
(ii) $f(z) = \begin{cases} 0.5, & z = 0, \\ 0.5, & 0 < z < 1. \end{cases}$

I-46. Let X be a uniform random variable on the interval $(-a, a)$ and Y a random variable independent of X, with distribution function $F(y)$. Show that the distribution function of $G = X + Y$ is

$$G(z) = \frac{1}{2a} \int_{-a}^{a} F(z - x) \, dx.$$

I-47. Let X be uniformly distributed on the interval $(0, 1)$. Find the probability density function of:

(i) $Y = aX + b$; a, b constants $(a > 0)$;

(ii) $Z = 1/X$;

(iii) $W = g(X)$, where g is a continuous and monotone function of x in $(0, 1)$.

Answer:

(i) Uniform;

(ii) $f(z) = 1/z^2$, $1 < z < \infty$;

(iii) $f(w) = 1/|g'(w)|$.

In addition to Problems I-23 (John Smith, 1963) and I-30 (points of Pascal), it is worth giving here some other problems which are of some historical interest, and show the difficulty in solving probability problems encountered by even well-known figures in the history of mathematics.

I-48. *Luca dal Borgo or Paccioli, 1494.* Each of three players A, B, and C has probability $1/3$ of hitting a target better than the other two. At each round of the game each shoots at the target once. The winner is the first player to win six rounds. They bet 10 ducats each. At a certain point, however, when A has 4 victories, B has 3 victories, and C has 2 victories, they have to terminate the game. How should they split their bet? (cf. Problem 30).

I-49. *Huyghens,* 1657.* A and B play a game in which each of them throws two ordinary dice. A wins if he scores 6 and B wins if he scores 7. First, A throws the dice once, then B throws the dice twice, and they continue alternating after each throw until one of them wins. What is the probability of winning for each of them?

Note: Huyghens and Spinoza gave the same answer:

$$P(A)/P(B) = 10355/1227 \approx 0.84.$$

(cf. Exercise 67).

I-50. *Huyghens, 1657.* A deck of forty cards consists of 10 red, 10 black, 10 blue, and 10 green cards. Four cards are drawn at random and without replacement. What is the probability that all the colors will show up?

(Huyghens gave the answer $1,000/9,129 = 0.109$; is it correct?)

I-51. *Huyghens, 1657.* An urn contains 8 black balls and 4 white balls. Each of three players A, B, and C (and in that order) draws a ball without replace-

---

* He is the author of the first book on probability, *De Ratiociniis in Ludo Aleae*, 1654.

ment and the first one to draw a white ball wins. What is the probability of winning for each of them?

Hint: *A* can win on the first, fourth or seventh draw, etc.

I-52.* *Montmort*, 1708. Three players *A*, *B*, and *C* play the following game. They play three rounds for each of which, each of them has probability 1/3 of winning. *A* is the winner of the game if he wins a round before *B* or *C* wins two rounds; *B* is the winner if he wins two rounds before player *A* wins one round, or *C* wins two rounds; *C* wins like *B*. What is the probability of winning for each of them?

(Montmort gave $P(A) = 15/27$, $P(B) = P(C) = 5/27$).

I-53.* *Daniel Bernoulli*, 1760. A container has $2N$ cards, numbered in such a way that each of the numbers 1, 2, ..., N appears twice. $2N - n$ cards are drawn at random. What is the expected number of pairs of cards remaining in the container?

(Bernoulli found $n(n - 1)/(4N - 2)$).

I-54. *Euler*, 1763. *N* lottery tickets are numbered 1 to *N*. *n* tickets are selected at random. What is the probability that *r* consecutive numbers are selected ($r = 1, 2, ..., n$)?

Example: For $N = 10$, $n = 4$, $r = 4$, the probability is $7 / \binom{10}{4} = 1/30$.

I-55.* *Condorcet*, 1785. The probability of event *A* is *p*. What is the probability that:

(a) in *n* trials event *A* occurs *k* times consecutively?

(b) in *n* trials event *A* occurs *k* times consecutively and then A^c also occurs *k* times consecutively?

(c) in $n + m$ trials event *A* occurs *n* times given that in $r + k$ trials it occurred *r* times?

Condorcet gave the following answer for (c):

$$(r + k + 1)! \, (r + n)! \, (k + m)!/r! \, k(r + k + m + n + 1)!.$$

Laplace and G. Polya disagreed with this solution. What do you say?

I-56.* *Laplace*, 1812. In a lottery with *N* tickets, *n* numbers are chosen as winners. What is the probability *p* that in *k* lottery games all the *N* numbers will appear?

(Laplace for $N = 90$, $n = 5$, and $k = 85$ gave $p = 1/2$).

Complements and Problems

1. Multivariate Distributions

1.1. The density of the impact point (X, Y) of a circular target is given by

$$f(x, y) = c(R - \sqrt{x^2 + y^2}) \quad \text{for} \quad x^2 + y \le R^2.$$

Find, (a) the constant c, (b) the probability p that the impact point falls in a circle with radius $a < R$ centered at the origin.

Answer: (a) $c = 3/R^2$. (b) $p = 3a^2/R^2(1 - 2a/3R)$.

1.2. The random variables X, Y satisfy the linear relation

$$aX - bY = c,$$

where a, b, c are constants. Find, (a) the correlation $\rho(X, Y)$, and (b) the ratio σ_x/σ_y of the standard deviations.

Answer: (a) $\rho = \text{sg}(ab)$. (b) $\sigma_x/\sigma_y = |b/a|$.

1.3. The dispersion (variance–covariance) matrix of a three-dimensional normal distribution is

$$(\sigma_{ij}) = \begin{bmatrix} 5 & 2 & -2 \\ 2 & 6 & 3 \\ -2 & 3 & 8 \end{bmatrix}.$$

If $\mu_x = \mu_y = \mu_z = 0$, find the density $f(x, y, z)$ and its maximum.

Answer: $f(x, y, z) = 1/(2\pi\sqrt{230\pi}) \exp\{-\frac{1}{230}(39x^2 + 36y^2 + 26z^2$

$$- 44xy + 36xz - 38yz)\}$$

$$f_{\max} = f(0, 0, 0) = (2\pi\sqrt{230\pi})^{-1} = 0.00595.$$

1.4. The joint density of X and Y is given by

$$f(x, y) = ye^{-y(x+1)}, \qquad x > 0, \quad y > 0.$$

Find, (a) the marginal densities of X and Y, and (b) the conditional distribution function $F_X(x|y)$ of X given $Y = y$.

Answer: (a) $f_X(x) = 1/(x + 1)^2$, $f_Y(y) = e^{-y}$. (b) $1 - e^{-yx}$.

1.5. Suppose that the random variable X coincides with the random variable X_i with probability p_i, that is,

$$P[X = X_i] = p_i \qquad \text{and} \qquad E(X_i) = m_i, \qquad i = 1, 2, \dots .$$

Show that

$$V(X) = \sum_i p_i V(X_i) + V(M),$$

where $P[M = m_i] = p_i$ $(i = 1, 2, \dots)$.

1.6. The random vector (X, Y, Z) has density $f(x, y, z)$. X, Y, Z can only be observed simultaneously. An observation gave the values u, v, w without knowing the correspondence between them and X, Y, Z. Find the probability that $X = u$, $Y = v$, and $Z = w$.

Answer:

$$\frac{f(u, v, w)}{\{f(u, v, w) + f(u, w, v) + f(v, u, w) + f(v, w, u) + f(w, u, v) + f(w, v, u)\}}.$$

1.7. The joint distribution function $F(x, y)$ of the random variables X and Y is given by

$$F(x, y) = \sin x \sin y, \qquad 0 \le x \le \pi/2, \quad 0 \le y \le \pi/2.$$

Find, (a) the density function $f(x, y)$, and (b) the dispersion matrix.

Answer: (a) $f(x, y) = \cos x \cos y$. (b) $\sigma_x^2 = \sigma_y^2 = \pi - 3$, $\sigma_{xy} = 0$.

1.8. The normal density of a point $A = (X_1, X_2, X_3)$ is given by

$$f(x, y, z) = \frac{1}{(2\pi)^{3/2}\sigma_1\sigma_2\sigma_3} \exp\left\{-\frac{1}{2}\left(\frac{x^2}{\sigma_1^2} + \frac{y^2}{\sigma_2^2} + \frac{z^2}{\sigma_3^2}\right)\right\}.$$

What is the probability that A will fall into an ellipsoid, with principal semiaxes E_1, E_2, E_3 along the axes Ox, Oy, Oz, where E_1, E_2, E_3 are the probable errors of X, Y, Z, respectively, i.e.,

$$P[|X| < E_1] = P[|Y| < E_2] = P[|Z| < E_3] = 1/2.$$

Answer: $(1/2) - (\sqrt{2}E_0/\sqrt{\pi})e^{-1/2E_0^2}$, where E_0 is the probable error of $N(0, 1)$.

1.9. The vector (X, Y) has the uniform distribution in the ellipse

$$\frac{x^2}{a^2} + \frac{y^2}{b^2} \le 1.$$

Find:

(a) the marginal distributions of X and Y;

(b) the conditional distributions of X given that $Y = y$, and the conditional distributions of Y given that $X = x$;

(c) the $\text{Cov}(X, Y) = \sigma_{xy}$.

Are X, Y independent?

Answer: (a) $f_X(x) = 2\sqrt{1 - (x^2/a^2)}/\pi a$, $f_Y(y) = 2\sqrt{1 - (y^2/b^2)}/\pi b$ ($|x| < a$, $|y| < b$).

(b) $f_Y(y|x) = (2b\sqrt{1 - (x^2/a^2)})^{-1}$ for $|x| < a$, $|y| < b\sqrt{1 - (x^2/a^2)}$; $f_X(x|y)$ similar.

(c) $\sigma_{xy} = 0$ but they are dependent since $f(x, y) \neq f_X(x)f_Y(y)$.

1.10. The point (X, Y, Z) is uniformly distributed in the sphere $x^2 + y^2 + z^2 \leq R^2$. Find the density of Z and the conditional density $f(x, y|z)$. Generalize for an n-dimensional sphere.

Answer: $f_Z(z) = 3(R^2 - z^2)/4R^3$, $|z| < R$, $f(x, y|z) = 1/\pi(R^2 - z^2)$.

1.11.* Given $f_Y(y)$, $E[X|y]$, $V(X|y)$, find $E(X)$ and $V(X)$.

Answers:

$$E(X) = \int_{-\infty}^{\infty} E(X|y)f_Y(y)\,dy = E_y[E(X|Y = y)],$$

$$V(X) = E[V(X|Y)] + V[E(X|Y)] \qquad \text{(Exercise 328)}$$

$$= \int_{-\infty}^{\infty} V(X|y)f_Y(y)\,dy + \int_{-\infty}^{\infty} [E(X|y) - E(X)]^2 f_Y(y)\,dy.$$

1.12. The vector (X, Y) has normal density

$$f(x, y) = ce^{-4x^2 - 6xy - 9y^2}.$$

Find the densities $f_X(x)$, $f_Y(y)$, $f_X(x|y)$, $f_Y(y|x)$ and the constant c.

Answer:

$$c = \frac{3\sqrt{3}}{\pi}, \qquad f_X(x) = \frac{\sqrt{3}}{\sqrt{\pi}}\exp\{-3x^2\}, \qquad f_Y(y) = \frac{3\sqrt{3}}{2\sqrt{\pi}}\exp\left\{-\frac{27}{4}y^2\right\},$$

$$f_X(x|y) = \frac{2}{\sqrt{\pi}}e^{-(2x+1.5y)^2}, \qquad f_Y(y|x) = \frac{3}{\sqrt{\pi}}e^{-(3y+x)^2}.$$

Note that they are all normal.

1.13.* Suppose that $\mathbf{X}$ has a *spherically symmetric n-dimensional* distribution, i.e., with density $f(x_1^2 + \cdots + x_n^2)$. Prove that the density g of the "*generalized χ^2 variable*" $U = \mathbf{X'X}$ is given by

$$g(u) = \tfrac{1}{2}C_n f(u)u^{n/2-1},$$

where C_n is the "surface" area of the n-dimensional sphere, i.e.,

$$C_n = \frac{2\pi^{n/2}}{\Gamma(n/2)}.$$

Hint: Make use of spherical (polar) coordinates $r, \theta_1, \ldots, \theta_{n-1}$: $x_1 = r \sin \theta_1$,

$$x_k = r \sin \theta_k \prod_{i=1}^{k-1} \cos \theta_i, \qquad k = 1, \ldots, n-1, \qquad x_n = r \prod_{i=1}^{n-1} \cos \theta_i,$$

where $-\frac{1}{2}\pi < \theta_i \leq \frac{1}{2}\pi$ $(i = 1, \ldots, n-2$ and $-\pi < \theta_{n-1} \leq \pi)$. Prove that the Jacobian is $r \prod_{k=1}^{n-2} (r \cos^{n-k-1} \theta_k)$ and make use of $\int_{-\pi/2}^{\pi/2} \cos^{k-1} \theta \, d\theta = B(\frac{1}{2}, k/2)$.

1.14.* (Continuation). For $n = 2$, show that the polar coordinates (R, θ), where

$$X = R \cos \theta, \qquad Y = R \sin \theta,$$

are independent, θ is uniformly distributed on $[0, 2\pi)$, and R has density (cf. Exercise 221)

$$h(r) = 2\pi r f(r^2).$$

1.15.* (Continuation). Show that the correlation coefficient based on the single observation (X, Y), i.e.,

$$r_0 = \frac{XY}{X^2 + Y^2}$$

has density

$$f(r_0) = \frac{1}{\pi} \frac{1}{\sqrt{1 - r_0^2}}, \qquad -1 < r_0 < 1.$$

Hint: $r_0 = \sin 2\theta$ has the same distribution as under a spherical normal (centered at the origin). Hence conclude that $\tan 2\theta = r_0/\sqrt{1 - r_0^2}$ has the Cauchy distribution (t with one degree of freedom).

1.16.* If $X = (X_1, \ldots, X_n)'$ is $N(0, I_n)$ and X is partitioned into k subvectors $X_{(1)}, \ldots, X_{(k)}$ with $v_1, v_2, \ldots, v_k$ components, respectively, show that $(X'_{(1)}X_{(1)}, \ldots, X'_{(k)}X_{(k)})/|X|^2$ has the Dirichlet distribution (see (5.11)), with density

$$f(t_1, \ldots, t_{k-1}) = \frac{\Gamma\left(\dfrac{n}{2}\right)}{\prod_{i=1}^{k} \Gamma\left(\dfrac{v_k}{2}\right)} (1 - t_1 - \cdots - t_{k-1})^{(v_k/2)-1} \prod_{i=1}^{k-1} t_i^{(v_i/2)-1},$$

$$t_i \geq 0, \quad \Sigma t_i < 1.$$

Hence, for $v_1 = v_2 = \cdots = v_s = 1$, $v_{s+1} = n - s$ $(k = s + 1)$,

$$\frac{(|X_1|, |X_2|, \ldots, |X_k|)}{|X|}$$

has the density

$$f_0(v_1, \ldots, v_k) = \frac{\Gamma\left(\dfrac{n}{2}\right) 2^k}{\Gamma\left(\dfrac{n-k}{2}\right)\pi}(1 - v_1^2 - \cdots - v_k^2)^{((n-k)/2)-1},$$

$$v_i \geq 0, \quad \sum_{i=1}^{k} v_i^x < 1,$$

and the density of $(X_1, \ldots, X_k)/|X| \equiv (U_1, U_2, \ldots, U_k)$ is

$$\frac{f_0(u_1, \ldots, u_k)}{2^k} \quad \text{if} \quad \sum_{i=1}^{k} u_i^2 < 1.$$

This is the distribution of any k-dimensional subvector of an n-dimensional vector

$$U = (U_1, \ldots, U_n)$$

uniformly distributed on the (surface of the) unit sphere

$$x_1^2 + \cdots + x_n^2 = 1.$$

1.17.* (*Continuation*). In general, if X has a spherical distribution (see Problem 1.13), since this is invariant under rotations in n-space, the distribution of the direction $X/|X|$ is the same as that of U and is independent of its length $|X|$ (for any spherical distribution with $P[|X| = 0] = 0$).

Hence conclude that the results of Problem 1.16 hold if X is any spherically symmetric random vector.

1.18. Let X, Y be random variables with

$$E(X) = \mu, \quad E(Y) = \eta, \quad V(X) = \sigma_1^2, \quad V(Y) = \sigma_2^2, \quad \text{Cov}(X, Y) = \sigma_{12}.$$

Find an approximate formula for

$$E[g(X, Y)], V[g(X, Y)],$$

using the linear expansion of $g(x, y)$ around (μ, η) (up to second-order moments). Generalize for n random variables.

Answer:

$$E[g(X, Y)] \simeq g(\mu, \eta) + \frac{1}{2}\left[\frac{\partial^2 g}{\partial x^2}\sigma_1^2 + 2\frac{\partial^2 g}{\partial x \partial y}\sigma_{12} + \frac{\partial^2 g}{\partial y^2}\sigma_2^2\right],$$

$$V[g(X, Y)] \simeq \left(\frac{\partial g}{\partial x}\right)^2 \sigma_1^2 + 2\frac{\partial g}{\partial x}\frac{\partial g}{\partial y}\sigma_{12} + \left(\frac{\partial g}{\partial y}\right)^2 \sigma_2^2,$$

where the partial derivatives are evaluated at the point (μ, η).

Let $X = (X_1, \ldots, X_n)'$, $E(X) = (\mu_1, \ldots, \mu_n)' = \mu$, $\text{Cov}(X) = \Sigma = (\sigma_{ij})$, the covariance matrix of X (see (5.7)), and

$$\nabla g = \left(\frac{\partial g}{\partial x_1}, \ldots, \frac{\partial g}{\partial x_n}\right)', \qquad \frac{\partial^2 g}{\partial x_i \partial x_j} = g_{ij},$$

where the derivatives of $g(x_1, \ldots, x_n)$ are evaluated at $\boldsymbol{\mu}$. Then

$$E[g(X)] \simeq g(\boldsymbol{\mu}) + \frac{1}{2} \sum_{i=1}^{n} \sum_{j=1}^{n} g_{ij}\sigma_{ij},$$

$$V[g(X)] \simeq (\nabla g)' \Sigma (\nabla g).$$

1.19. X and Y are independent and identically distributed random variables with density given by

$$f(x) = \frac{2}{\pi\sqrt{1 - x^2}}, \qquad 0 \le x \le 1.$$

Using Problem 1.18 find the $E(Z)$ and $V(Z)$ where

$$Z = \arctan \frac{X}{Y}.$$

Answer:

$$E(Z) = \arctan \frac{E(X)}{E(Y)} \simeq \frac{\pi}{4}, \qquad V(Z) \simeq \frac{\pi^2 - 8}{16}.$$

1.20. The fundamental frequency of a chord is given by the formula

$$\Omega = \frac{1}{2} \sqrt{\frac{F}{mL}},$$

where F is the tension, m is the mass of the chord, and L is its length. If the mass is assumed to be constant and F, L are random variables with $E(F) = f$, $E(L) = l$, $V(F) = \sigma_f^2$, $V(L) = \sigma_l^2$, and $\mathrm{Cov}(F, L) = \sigma_{fl}$, find an approximation for the variance of Ω.

Answer:

$$V(\Omega) \simeq \frac{f}{16ml}\left[\frac{\sigma_f^2}{f^2} + \frac{\sigma_l^2}{l^2} - \frac{2\sigma_{fl}}{fl}\right] \qquad \text{(see Problem 1.13)}.$$

1.21. Let the vector (X, Y) have density $f(x, y)$. We define the complex variables

$$Z = X + iY, \qquad Z_t = Ze^{it},$$

and $Z_t = Ze^{it}$. Prove that, in order for all Z_t to have the same distribution, it is necessary that $f(x, y) = g(x^2 + y^2)$ for some density g (see Problem 1.13).

Hint: The distribution of Z_t has to be invariant under orthogonal transformations (rotations).

1.22. If $\mathbf{X}$ has density f as in Problem 1.13, show that

$$\mathbf{Y} = \boldsymbol{\mu} + A\mathbf{X} \qquad \text{(a nonsingular } n \times n \text{ constant matrix)}$$

has density of the form

$$c|\Lambda|^{-1/2} f((\mathbf{y} - \boldsymbol{\mu})' \Lambda^{-1} (\mathbf{y} - \boldsymbol{\mu})), \qquad \Lambda = AA',$$

and characteristic function of the form

$$e^{it'\mu}\psi(t'\Lambda t) \qquad \text{for some function } \psi.$$

1.23. Let $\rho_{ij} = \rho(X_i, X_j)$ be the correlation coefficient between X_i, X_j for i, $j = 1, \ldots, n$. For $n = 3$ and $\rho_{12} = \rho_{13} = \rho_{23} = \rho$, what are the possible values that ρ can take?

Answer: $-1/2 \le \rho \le 1$. Prove that the correlation matrix $P = (\rho_{ij})$ is nonnegative definite, i.e., $t'Pt \ge 0$ for all $t' = (t_1, \ldots, t_n)$.

1.24. If the p vector X is $N(\mu, \Lambda)$, show that the quadratic form

$$\chi^2 = (X - \mu)'\Lambda^{-1}(X - \mu)$$

has the χ_p^2 distribution, with p degrees of freedom, and density given by (cf. Problem 1.13)

$$f_p(x) = \frac{1}{2^{n/2}\Gamma(p/2)} x^{(p/2)-1} e^{-x/2}.$$

1.25.* The entropy H of a multivariate distribution with density $f(x)$ is defined by the formula

$$H(f) = -E[\log f(X)] = -\int \cdots \int f(x) \log f(x)\, dx.$$

Prove that for the normal distribution of Problem 1.24

$$H = \log[(2\pi e)^{p/2}|\Lambda|^{-1/2}].$$

1.26. If X is $N(\mu, \Lambda)$ with density f, find the density g of the random variable

$$Y = f(X).$$

Hint: See Problem 1.24.

$$g(y) = 2\sqrt{c}\left[\log \frac{1}{cy^2}\right]^{(p/2)-1}, \qquad c = (2\pi)^p|\Lambda|.$$

1.27. If $X = (X_1, \ldots, X_n)$ is uniformly distributed in the simplex T:

$$x_1 + x_2 + \cdots + x_n \le 1, \qquad x_i \ge 0, \qquad i = 1, \ldots, n,$$

show that the density of X is

$$f(x) = n! \qquad \text{for } x \in T.$$

Show that the $(n - 1)$-dimensional marginals are also special cases of the Dirichlet distribution (5.11).

2. Generating Functions

2.1. Two persons shoot at a target, each has n shots. Using generating functions, find the probability p that they have the same number of successes, given that each of them has a probability 0.5 of success (hitting the target) at each shot.

Answer: The coefficient of the constant term in the expansion of the generating function

$$G(t) = \frac{1}{2^n}(t + 1)^n \frac{1}{2^n}\left(1 + \frac{1}{t}\right)^n = \frac{(1 + u)^{2n}}{4^n u^n},$$

that is,

$$p = \frac{1}{4^n}\binom{2n}{n}.$$

2.2.* To obtain the title of champion in chess, an opponent has to win at least 12.5 points out of the maximum of 24. In the case of a tie (12:12), the defending champion maintains the title. The probability that each of the two players wins a game is the same, and equals half the probability of a tie. Find:

(a) the probability p_{ch} that the defending champion will maintain his title, as well as the probability p_{op} that the opponent will become the champion;

(b) the probability p that 20 games will be played in the contest.

Answer: (a) The probability p_{ch} equals the sum of the coefficients of the nonnegative powers of t in the expansion of

$$G(t) = \left(\frac{1}{4}t + \frac{1}{4t} + \frac{1}{2}\right)^{24} = \frac{(1 + t)^{48}}{4^{48}t^{24}},$$

$$p_{ch} = \frac{1}{4^{24}}\sum_{k=24}^{48}\binom{48}{k} = \frac{1}{2}\frac{1}{4^{24}}\left[2^{48} + \binom{48}{24}\right] = 0.5577, \qquad p_{op} = 1 - p_{ch}.$$

(b) The probability of the complementary event equals the sum of the coefficients of t, with exponents between -4 up to 3 in the expansion of

$$G(t) = \frac{1}{4^{20}}\frac{(1 + t)^{20}}{t^{20}}; \qquad p = 1 - \frac{1}{4^{20}}\sum_{k=16}^{23}\binom{40}{k} = 0.22.$$

2.3. In a lottery, tickets are numbered from 000000 to 999999 and they are all equally probable. What is the probability p that the sum of the digits of a chosen number be 21?

Answer: The probability p is the coefficient of t^{21} in the expansion of

$$G(t) = \frac{1}{10^6}(1 + t + t^2 + \cdots + t^9)^6 = \frac{1}{10^6}\frac{(1 - t^{10})^6}{(1 - t)^6}$$

$$= \frac{1}{10^6}\left\{1 - \binom{6}{1}t^{10} + \binom{6}{2}t^{20} - \cdots\right\}\left\{1 + \binom{6}{5} + \binom{7}{5}t^2 + \cdots\right\},$$

where we made use of the equality

$$\frac{1}{(1-t)^n} = 1 + \binom{n}{n-1}t + \binom{n+1}{n-1}t^2 + \cdots.$$

So

$$p = \frac{1}{10^6}\left\{\binom{26}{5} - \binom{6}{1}\binom{16}{5} + \binom{6}{2}\binom{6}{5}\right\} = 0.04.$$

2.4. Prove that the Pascal distribution, for $r \to \infty$ and $rq \to \lambda$, converges to the Poisson with parameter λ.

2.5. Let X, Y, Z be independent random variables, each taking the values $1, 2, \ldots, n$ with the same probability (uniformly). Find $P[X + Y = 2Z]$ from their joint generating function $G(t_1, t_2, t_3)$.

Answer: Equal to the constant term $1/2n$ of the generating function G_w of $W = X + Y - 2Z$.

$$G_w(s) = G(s, s, s^{-2}) = \frac{(1 - s^n)^2(1 - s^{2n})}{n^3(1 - s)^2(1 - s^2)s^{2n-2}}.$$

2.6. In Exercise 315, let

$$q_{n,k} = P[S_n \le k].$$

Find the generating function $Q(t)$ of $q_{n,k}$ for a given n and hence $q_{n,k}$.

Answer: $1 - P(t) = (1 - t)Q(t)$ (where $P(\cdot)$ is the generating function of S_n).

So

$$q_{n,k} = 1 - \frac{1}{m^n} \sum_{j<k/m} (-1)^j \binom{n}{j}\binom{k+n-mj}{n}.$$

2.7. (*Continuation*). Suppose k and m tend to infinity so that $k/m \to x$. Find the limit of $q_{n,k}$.

Answer:

$$1 - \frac{1}{n!} \sum_{j<x} (-1)^j \binom{n}{j}(x - j)^n = 1 - F(x),$$

where F is the distribution function of the sum of n independent random variables from the uniform distribution $(0, 1)$ (see Exercise 314).

2.8. Find the densities of the distributions with characteristic functions

$$\varphi_1(t) = \frac{1+it}{1+t^2}, \qquad \varphi_2(t) = \frac{1-it}{1+t^2}.$$

Answer: $f_1(x) = e^{-x}, x > 0, f_2(x) = e^x, x < 0.$

2.9. Find the moments of the Laplace distribution with characteristic function

$$\varphi(t) = \frac{1}{1+t^2}.$$

Answer: Using the Taylor series expansion

$$\frac{1}{1 + t^2} = \sum_{k=0}^{\infty} (it)^{2k},$$

conclude that

$$\mu_k' = \begin{cases} k! & \text{for even } k, \\ 0 & \text{for odd } k. \end{cases}$$

2.10. Determine the distribution with characteristic function

$$\varphi(t) = \frac{1}{2e^{-it} - 1}.$$

Answer: Discrete random variable with probability function

$$P[X = k] = 2^{-k}, \qquad k = 1, 2, \dots.$$

Make use of the expansion of $\varphi(t)$ in powers of $\frac{1}{2}e^{it}$ and the expression of the Dirac function δ

$$\delta(x) = \frac{1}{2\pi} \int_{-\infty}^{\infty} e^{itx} \, dt.$$

2.11. Find the characteristic function of $Y = aF(X) + b$, where X is a continuous random variable and F is its distribution function.

Hint: Y is uniform on (a, b); see Exercise 227.

2.12. A vector $\xi = (X_1, \dots, X_n)$ follows the n-dimensional normal distribution with $E(X_i) = a$, $V(X_i) = \sigma^2$ $(i = 1, \dots, n)$ and covariances

$$\text{Cov}(X_i, X_{i+1}) = \rho\sigma^2, \qquad i = 1, 2, \dots, n - 1,$$

$$\text{Cov}(X_i, X_j) = 0 \qquad \text{for} \quad |i - j| > 1.$$

Find the characteristic function of ξ.

Answer:

$$\varphi(t_1, \dots, t_n) = \exp\left\{ ia \sum_{k=1}^{n} t_k - \frac{\sigma^2}{2} \sum_{k=1}^{n} t_k^2 - \rho\sigma^2 \sum_{k=1}^{n-1} t_k t_{k+1} \right\}.$$

2.13. Let $\varphi(t)$ be the characteristic function of a random variable X with $E(X^2) = \mu_2'$.

(a) Prove that

$$\psi(t) = -\frac{1}{\mu_2'} \varphi''(t)$$

is a characteristic function.

(b) Conclude that

$$\psi_0(t) = (1 - t^2)e^{-t^2/2}$$

is a characteristic function.

Answer: (a) $\psi(t)$ is the characteristic function of the random variable with distribution function

$$G(x) = \frac{1}{\mu_2'} \int_{-\infty}^{x} u^2 \, dF(u).$$

(b) Application of (a) with X normal $N(0, 1)$.

2.14. Show that

$$a(t) = \begin{cases} 1 - (|t|/\pi) & \text{for } |t| < \pi, \\ 0 & \text{for } |t| > \pi, \end{cases}$$

is the characteristic function of a random variable X, and determine the corresponding distribution.

Answer: Since $|a(t)|$ is integrable, X is continuous with density

$$f(x) = \frac{1}{2\pi} \int_{-\pi}^{\pi} e^{-itx} \left(1 - \frac{|t|}{\pi} \right) dt = \frac{1 - \cos(\pi x)}{\pi^2 x^2}.$$

2.15. Let X_1, X_2, X_3 be independent $N(0, 1)$. Find the characteristic function of the pair Z_1, Z_2, where

$$Z_1 = X_2 X_3, \qquad Z_2 = X_1 X_3.$$

Hence conclude the characteristic function of Z_1. Are Z_1, Z_2 independent?
Answer:

$$\varphi(t_1, t_2) = (1 + t_1^2 + t_2^2)^{-1/2},$$

$$\varphi_{Z_1}(t) = \varphi(t, 0) = (1 + t^2)^{-1/2} = \varphi_{Z_2}(t).$$

No, because $\varphi(t_1, t_2) \neq \varphi_{Z_1}(t_1)\varphi_{Z_2}(t_2)$.

2.16. Find the distributions corresponding to the characteristic functions

$$\varphi_1(t) = \frac{1}{\cosh t}, \qquad \varphi_2(t) = \frac{1}{\sinh t}, \qquad \varphi_3(t) = \frac{1}{\cosh^2 t}.$$

Answer:

$$f_1(x) = \frac{1}{2 \cosh \dfrac{\pi x}{2}}, \qquad f_2(x) = \frac{\pi}{4 \cosh^2 \dfrac{\pi x}{2}}, \qquad f_3(x) = \frac{1}{2 \sinh \dfrac{\pi x}{2}}.$$

2.17. *Bivariate negative binomial.* Show that

$$P(s_1, s_2) = p_0^r(1 - p_1 s_1 - p_2 s_2)^{-r},$$

with positive parameters and $p_0 + p_1 + p_2 = 1$ is the probability generating function of a pair (X, Y), so that the marginal distributions of X and Y and $X + Y$ are negative binomials.

2.18. *Problems related to a game of billiards.* Let a "round" consist of a sequence of Bernoulli trials before the first failure. Find the generating function and the probability function of the total number S_k of successes in k rounds.

Answer: Pascal distribution.

2.19. (*Continuation*). Let G be the number of successive rounds up to the nth success (i.e., the nth success occurs during the Gth round). Find the distribution of G and the

$$E(G), \ V(G).$$

Answer:

$$P[G = k] = \sum_{j=0}^{n-1} P[S_{k-1} = j]P[X_k \geq n - k]$$

$$= \sum_{j=0}^{n-1} q^{k-1}p^j \binom{k+j-2}{j}p^{n-j}\binom{k+n-2}{n-1},$$

$$E(G) = 1 + \frac{nq}{p}, \qquad V(G) = \frac{nq}{p^2}.$$

2.20.* (*Continuation*). Consider two independent sequences of Bernoulli variables with parameters p_1 and p_2, respectively, or two players with "skills" p_1 and p_2, respectively. Prove that each player will need the same number of rounds (trials) until the Nth success (a tie) with probability

$$(p_1p_2)^N \sum_{v=1}^{\infty} \binom{N+v-2}{v-1}(q_1q_2)^{v-1}$$

$$= (p_1p_2)^N(1 - q_1q_2)^{1-2N} \sum_{k=0}^{N-1} \binom{N-1}{k}(q_1q_2)^k.$$

2.21. Show that the characteristic function of the β density (2.10) is

$$\frac{\Gamma(p+q)}{\Gamma(p)} \sum_{j=0}^{\infty} \frac{\Gamma(p+j)(it)^j}{\Gamma(p+q+j)\Gamma(j+1)}.$$

2.22.* Show that the characteristic function of the density in Problem 1.13 is of the form $\varphi(t_1^2 + t_2^2 + \cdots + t_n^2)$ (see Problem 1.21). If

$$\varphi(t_1^2 + \cdots + t_n^2) = \prod_{i=1}^{n} \varphi(t_i^2), \qquad (*)$$

that is, $X_1, X_2, \ldots, X_n$ are independent, prove that $\mathbf{X}$ is $N(0, I)$. In other words, *a spherically symmetric random vector has independent components if and only if it is multivariate normal.*

Hint: Equation $(*)$ is known as Hamel's equation and its only continuous solution is

$$\varphi(t) = e^{ct}$$

for some constant c (which for a characteristic function must be < 0).

3. Transformation of Random Variables

3.1. Find the density of

$$Y = +\sqrt{|X|},$$

where $X \sim N(0, 1)$.
Answer:

$$f_Y(y) = \frac{4y}{\sqrt{2\pi}} \exp\left\{-\frac{y^4}{2}\right\}, \qquad y \geq 0.$$

3.2. X has a uniform distribution on $(0, 1)$. Find the density of the random variable Y defined by

$$X = \frac{1}{2}\left\{1 + \frac{2}{\sqrt{2\pi}} \int_0^Y \exp\left\{-\frac{t^2}{2}\right\} dt\right\}.$$

Answer: Y is $N(0, 1)$ (see Exercise 227).

3.3. The random variables X and Y are related by

$$X = \int_{-\infty}^Y f(t)\, dt,$$

where $f(t) \geq 0$ satisfies the

$$\int_{-\infty}^{\infty} f(t)\, dt = 1.$$

Prove that if X is uniform on $(0, 1)$, then $f(t)$ is the density of Y (see Exercise 227).

3.4. Suppose X and Y are independent random variables. Prove that the product XY has the same distribution if:
(a) X and Y are $N(0, 1)$; or
(b) X is $N(0, 1)$ and Y has density

$$f(y) = ye^{-y^2/2}, \qquad y \geq 0 \qquad (\chi_2 \text{ density; see } (2.11)).$$

3.5. The roots of $x^2 + Yx + Z = 0$ follow the normal distribution $(-1, 1)$. Find the density of the coefficients Y and Z (cf. Exercise 222).
Answer:

$$f_Y(y) = \tfrac{1}{4}(2 - |y|) \qquad \text{for} \quad |y| \leq 2,$$
$$f_Z(z) = -\tfrac{1}{2} \log |z| \qquad \text{for} \quad |z| \leq 1.$$

3.6. Let $f(x, y, z)$ be the density of a point (X, Y, Z) in the three-dimensional space and

$$R^2 = X^2 + Y^2 + Z^2, \qquad \theta = \arctan\sin\frac{Y}{R}.$$

What is the density of the pair (R, θ)?

Answer: $f(r, \theta) = r^2 \cos \theta \int_0^{2\pi} f(r \cos \theta \cos \varphi, r \cos \theta \sin \varphi, r \sin \theta) \, d\varphi$ (cf. Problem 1.11).

3.7. X and Y are independent normal $N(0, \sigma^2)$. Find the distribution of (a) $Z = X/|Y|$; (b) $W = |X|/|Y|$.
Answer: (a) Cauchy Distribution (cf. Exercise 220). (b) $f(w) = (2/\pi) \times [1/(1 + w^2)]$ (folded Cauchy).

3.8. X and Y are independent random variables with uniform distribution on $(0, a)$. Find the distribution of $Z = X/Y$ and examine the existence of moments of Z.
Answer:

$$f(z) = \begin{cases} \frac{1}{2}, & z \le 1, \\ \dfrac{1}{2z^2}, & z \ge 1. \end{cases}$$

There are no ordinary moments.

3.9. The joint density of X, Y is

$$f(x, y) = e^{-y} \quad \text{for} \quad 0 \le x \le y < \infty.$$

(a) Find the marginal distributions of X, Y and the conditional distribution of Y for given X.
(b) Of the variables X, Y, $Y - X$, and X/Y which are pairwise independent?
(c) In the conditional distribution of Y, given $X = x$, determine the interval $(y, y + a)$ so that

$$Q(a) = P[y \le Y \le y + a]$$

can be maximized. These intervals with variable x define a zone B. What is the probability of the zone, i.e.,

$$P[(X, Y) \in B] = P(B)?$$

Answer: (a)

$$f_X(x) = \int_x^\infty e^{-y} \, dy = e^{-x}, \qquad f_Y(y) = y e^{-y},$$

$$f_Y(y|x) = e^{x-y}.$$

(b)

$$(X, Y - X) \qquad \text{and} \qquad (Y, X/Y).$$

(c)

$$Q(a) = \sup_{u \ge x} \int_u^{u+a} e^{x-y} \, dy = \sup_{u \ge x} e^{x-u}(1 - e^{-a}) \qquad \text{(for } u = x\text{)}.$$

$P(B) = Q(a)$ as also concluded from the independence of X and $Y - X$.

3.10.* X, Y are independent exponentials each with density

$$f(t) = \theta e^{-\theta t}, \qquad t \geq 0. \tag{1}$$

(a) Find the joint conditional density of X, Y, given that $X \leq Y$ (ordered sample of two observations).

(b) Let $X_1 \leq X_2 \leq X_3 \leq X_4$ be an ordered sample from (1). What is the density f_1 of (X_1, X_2) and what is the conditional density f_2 of (X_3, X_4) for a given (X_1, X_2)? What do you conclude?

Answer: (a)

$$f(x, y) = 2\theta^2 e^{-\theta(x+y)},$$

$$f_Y(y|x) = \theta e^{-\theta(y-x)}, \qquad 0 < x < y \qquad \text{(see Exercise 238)},$$

i.e., $Y - X$, given that $X = x$, has the density (1).

$$m(x) = E[Y|X = x] = \int_x^\infty y f_Y(y|x) \, dy = x + \frac{1}{\theta}.$$

This also follows from the fact that $Y - X$, given $X = x$, follows (1), whose mean is $1/\theta$.

(b)

$$f_1(x_1, x_2) = 12\theta^2 e^{-\theta(x_1 + 3x_2)},$$

$$f_2(x_3, x_4) = 2\theta^2 e^{-\theta(x_3 + x_4 - 2x_2)}, \qquad 0 < x_1 < x_2.$$

We observe that, for $X_2 = x_2$, the pair $(X_3 - X_2, X_4 - X_2)$ is distributed as (X, Y) in (a), i.e., the distribution is not affected by the knowledge that X exceeds a given value, because of the "lack of memory" of the exponential.

3.11.* If the random variables X_i are independent normal $N(a + bt_i, \sigma^2)$ ($i = 1, \ldots, n$), where t_i are constants with $\sum_{i=1}^n t_i = 0$, find:
  (a) the joint distribution of $X_1, \ldots, X_n$;
  (b) the joint distribution of

$$\bar{X} = \frac{1}{n} \sum_{i=1}^n X_i \quad \text{and} \quad b = \sum_{i=1}^n t_i X_i \Big/ \sum_{i=1}^n t_i^2.$$

Answer: $\bar{X}$ is $N(a, \sigma^2/n)$ and b is also $N(b, \sigma^2/\sum_{i=1}^n t_i^2)$.

3.12.* (Continuation). By a proper orthogonal transformation of $\mathbf{X} = (X_1, \ldots, X_n)'$, say $\mathbf{Y} = \mathbf{0X}$ ($\mathbf{0}$ $n \times n$ orthogonal matrix), prove that the random variables $\bar{X}$, b and $\sum_{i=1}^n (X_i - \bar{X} - bt_i)^2$ are completely independent. Conclude that $\sum_{i=1}^n (X_i - \bar{X} - bt_i)^2/\sigma^2$ has the χ^2 distribution with $n - 2$ degrees of freedom.

3.13. X and Y are jointly normal with $E(X) = E(Y) = 0$, $V(X) = V(Y) = 1$, and $\rho(X, Y) = \rho$, $|\rho| < 1$. Consider the polar coordinates $R = \sqrt{X^2 + Y^2}$

and $\theta = \tan^{-1}(Y/X)$. Show that θ has density (cf. Exercises 220, 221)

$$f(\theta) = \frac{\sqrt{1 - \rho^2}}{2\pi(1 - \rho \sin 2\theta)}, \qquad 0 < \theta < 2\pi.$$

3.14. *An elliptically symmetric t distribution.* The random vector $\mathbf{X} = (X_1, \ldots, X_k)$ has the density

$$f(\mathbf{x}) = \frac{\Gamma((k + n)/2)}{\Gamma(k/2)(k\pi)^{k/2}} [1 + (\mathbf{x} - \boldsymbol{\mu})'\Lambda^{-1}(\mathbf{x} - \boldsymbol{\mu})]^{-(n+k)/2}.$$

If $\Lambda = I$ and $k = 1$ this is the Student distribution with n degrees of freedom. If $\Lambda = I_k$, then X has a spherically symmetric t distribution (see Problem 1.13). Show that:

(a) $E(\mathbf{X}) = \boldsymbol{\mu}$.
(b) $\mathbf{X}$ has the same distribution as

$$\boldsymbol{\mu} + \Lambda^{1/2}\mathbf{Z}/\chi_n^2,$$

where $\mathbf{Z}$ is $N(0, I_k)$ independent of the χ_n^2. Hence conclude that (cf. Problem 1.24)

$$F = \frac{1}{k\chi_n^2}(\mathbf{X} - \boldsymbol{\mu})'\Lambda^{-1}(\mathbf{X} - \boldsymbol{\mu}) \quad \text{is } F_{k,n}.$$

(c) As $n \to \infty$,

$$\mathbf{X} \xrightarrow{L} \mathbf{Z}.$$

3.15. Let $f(x)$ denote the density of a random variable X. If $f(\cdot)$ is monotone and bounded, then $f(X)$ has a uniform distribution if and only if X has the (negative) exponential distribution.
    Hint: See (7.1) or Exercise 227.

4. Convergence of Random Variables

4.1. A Geiger–Müller counter registers a particle emitted by a radioactive source with probability 10^{-4}.
    (a) Assuming that during a certain period, the source emitted 3×10^4, what is the probability that more than two particles will be registered.
    (b) How many particles have to be emitted so that at least four particles will be registered with probability 0.99?
    Answer: (a) $1 - e^{-\lambda}[1 - \lambda - (\lambda^2/2)]$ with $\lambda = 3 \times 10^4 \times 10^{-4} = 3$.
    (b)

$$e^{-x}\left[1 + x + \frac{x^2}{2} + \frac{x^3}{6}\right] \le 0.01, \qquad x \simeq 10.7, \quad n = x \cdot 10^4 \simeq 107{,}000.$$

4.2. In some of the countries of western Europe, from the seventeenth century up to World War I, the following state lottery was played: 5 numbers out of 90 were drawn and a player had the right to bet, say, an amount a, on one or more numbers; if all the numbers he had chosen were among the 5 numbers drawn, then he would win an amount determined as follows: $15a$ if he had bet on one number, $270a$ if he had bet on 2 numbers, $500a$ if he had bet on 3 numbers, $75,000a$ if he had bet on 4 numbers, and $1,000,000a$ if he had bet on 5 numbers.

(a) What is the expected gain of a player who is betting on 1, 2, 3, 4, and 5 numbers?

(b) Suppose that 100,000 players bet on 3 numbers; what is the probability that at least 10 players win?

Answer: (a) Let p_k be the probability that a player who bets on k numbers will win, and that E_k will be his expected gain. Then

$$p_k = \frac{\binom{90 - k}{5 - k}}{\binom{90}{5}}; \quad p_1 = \frac{1}{18}, \quad p_2 = \frac{2}{801}, \quad p_3 = \frac{1}{11,748},$$

$$p_4 = \frac{1}{511,038}, \quad p_5 = \frac{1}{43,949,268},$$

$$E_1 = 15a\frac{1}{18} - a = -\frac{1}{6}a, \quad E_2 = -\frac{29}{89}a \approx -\frac{1}{3}a, \quad \text{etc.}$$

(b) 0.24.

4.3.* Making use of the Stirling formula prove that, as $\lambda \to \infty$ and for a constant k, we have the local limit theorem for the approximation of the Poisson probability function by the normal

$$\sqrt{\lambda}\left|\frac{\lambda^k}{k!}e^{-\lambda} - \frac{1}{\sqrt{2\pi\lambda}}\exp\left\{-\frac{1}{2\lambda}(k - \lambda)^2\right\}\right| \to 0.$$

4.4.* If $X_n \xrightarrow{P} c$ and f is continuous at c, then $f(X_n) \xrightarrow{P} f(c)$.

4.5.* Suppose $\{X_n, n \geq 1\}$ is a sequence of random variables with $E(X_n) = \theta$ and $V(X_n) = \sigma^2(\theta)$ where the parameter θ takes values in some interval. Prove that if $\sigma_n^2(\theta) \to 0$, then

$$E[g(X_n)] \to g(\theta)$$

for any bounded and continuous function g.

Answer: Examine the

$$\int_{-\infty}^{\infty} |g(x) - g(\theta)| \, dF_n(x)$$

for $|x - \theta| < \delta$ where $|g(x) - g(\theta)| < \varepsilon$, and for $|x - \theta| \geq \delta$ where $|g(x) - g(\theta)| \leq M < \infty$, and make use of the Chebyshev inequality ($F_n(x)$ denotes the distribution function of X_n).

4.6.* (*Continuation*). If F_n is the "binomial" with

$$\sigma_n^2(\theta) = \frac{\theta(1 - \theta)}{n},$$

and g is defined in $[0, 1]$, then the Bernstein polynomial

$$\sum_{k=0}^{n} g\left(\frac{k}{n}\right)\binom{n}{k}\theta^k(1 - \theta)^{n-k} \xrightarrow[n \to \infty]{} g(\theta) \tag{1}$$

uniformly in $0 \leq \theta \leq 1$. So we have a simple proof of the *Weierstrass Approximation Theorem*, according to which such a function g can be uniformly approximated by a polynomial. Moreover, (1) results in approximating polynomials.

4.7.* (*Continuation*). If nX_n is Poisson with parameter $n\theta$, then

$$e^{-n\theta}\sum_{k=0}^{\infty} g\left(\frac{k}{n}\right)\frac{(n\theta)^k}{k!} \to g(\theta) \tag{2}$$

and the convergence is uniform in every finite interval of values of θ. Equation (2) also holds true for nonintegral values of n.

4.8. Suppose X_1, X_2, ... is a sequence of independent normal random variables with $E(X_i) = 0$ and $V(X_i) = 1$ and let

$$\xi_n = \sqrt{n}\,\frac{X_1 + \cdots + X_n}{X_1^2 + \cdots + X_n^2}, \qquad \zeta_n = \frac{X_1 + \cdots + X_n}{\sqrt{X_1^2 + \cdots + X_n^2}}.$$

Prove that ξ_n and $\zeta_n \to N(0, 1)$.
    Hint: $X_1^2 + \cdots + X_n^2 \xrightarrow{P} n$; use (3.4)(i).

4.9. (*Continuation*). Find the asymptotic ($n \to \infty$) distribution of

$$\chi_n^2 = \sum_{k=1}^{n} X_k^2 \qquad \text{and} \qquad \xi_n = \frac{nX_{n+1}}{\chi_n^2}.$$

Answer: $(\chi_n^2 - n)/\sqrt{2n}$ and ξ_n are asymptotically $N(0, 1)$.

4.10. n numbers are chosen at random from the integers $1, 2, \ldots, N$, say $k_1, k_2, \ldots, k_n$. We set $X_i = 0$ if $k_i = 0$ modulo 3, $X_i = 1$ if $k_i = 1$ modulo 3, and $X_i = -1$ if $k_i = 2$ modulo 3. Set $S_n = \sum_{i=1}^n X_i$. Prove that when n and $N \to \infty$ so that $n/N \to c > 0$, $S_n/\sqrt{n} \xrightarrow{L} N(0, 2/3)$.

4.11.* Suppose $S_N = X_1 + X_2 + \cdots X_N$ is the sum of a random number N of random variables X_i, where X_i and N are independent, $|X_i| < c$, $E(X_i) = \mu$, $V(X_i) = \sigma^2$, $E(N) = n$, and $V(N) \leq n^{1-\varepsilon}$, $\varepsilon > 0$. Make use of Exercise 328 and the CLT, and prove that as $n \to \infty$

$$\frac{S_N - n\mu}{\sigma\sqrt{n}} \to N(0, 1).$$

4.12. Let $\{X_n, n \geq 1\}$ be a sequence of independent random variables and

$$P\left[X_n = \pm\frac{1}{2^n}\right] = \frac{1}{2}, \qquad n = 1, 2, \ldots.$$

Prove that the sum

$$S_n = X_1 + \ldots + X_n$$

(a) converges in law, and (b) find its asymptotic distribution.
Answer: (a) The characteristic function of S_n

$$\varphi_n(t) = \prod_{k=1}^{n} \cos\left(\frac{t}{2^k}\right),$$

because $\cos(t/2^k) - 1$ behaves in the same way as $-t^2/2^{2k+1}$.

(b)

$$\varphi_n(t) = \frac{\sin t}{t} \frac{t/2^n}{\sin(t/2^n)} \xrightarrow[n\to\infty]{} \frac{\sin t}{t},$$

i.e., $\varphi_n(t)$ converges to the characteristic function of the uniform in $[-1, 1]$.

4.13. If a sequence of normal random variable(s) X_n converges in law to the random variable X, then X is also normal or degenerate.
Answer: The characteristic function of X_n, $\varphi_n(t) = \exp(i\mu_n t - \frac{1}{2}\sigma_n^2 t^2) \to \varphi_X(t)$ uniformly for $|t| < \delta, \delta > 0$. So

$$|\varphi_n(t)|^2 = e^{-\sigma_n^2 t} \to |\varphi_X(t)|^2,$$

where, if $\{\sigma_n^2\}$ is bounded, then $\sigma_n^2 \to \sigma^2 \geq 0$; hence

$$e^{i\mu_n t} = \varphi_n(t)e^{(1/2)\sigma_n^2 t^2} \to \varphi(t)e^{(1/2)\sigma^2 t^2}.$$

Therefore μ_n converges, say, to μ and finally

$$\varphi_n(t) \to e^{i\mu t - (1/2)\sigma^2 t^2}.$$

If $\lim \sigma_n^2 = \sigma^2 = 0$, then X degenerates (shrinks) to the point μ.

4.14. Prove that the negative binomial

$$P[X = k] = \binom{r + k - 1}{k} p^r q^k, \qquad k = 0, 1, 2, \ldots, \tag{1}$$

with $r \to \infty$ and $q \to 0$ so that $rq \to \lambda > 0$, converges to a Poisson with parameter λ.

Answer: The generating function of (1),

$$\frac{(1-q)^r}{(1-qs)^r} \xrightarrow[rq\to\lambda]{} e^{\lambda(s-1)}.$$

4.15. If the stochastic sequence $\{X_n\}$ converges in probability to the random variable X, where $P[X \neq 0] = 1$, then the sequence $\{1/X_n\} \xrightarrow{P} 1/X$.

4.16. Consider the sequence $\{X_n\}$ *of independent random variables with*

$$P[X_n = \pm a_n] = 1/2, \qquad n = 1, 2, \ldots .$$

If the series Σa_n^2 converges, prove that the sequence of partial sums

$$S_n = X_1 + \cdots + X_n,$$

also converges in each of the four modes of convergence (see Chapter 8).

Answer: The Chebyshev inequality for $|S_n - S_m|$ and the Cauchy criterion imply the convergence of S_n. It should also be noted that the X_n are uniformly bounded, since $a_n \to 0$, and then the modes of convergence are equivalent.

4.17. Let $\{X_k\}$ be a sequence of random variable(s) with

$$P[X_k = \pm k^a] = 1/2.$$

(a) Find the values of a for which

$$\bar{X}_n = \frac{1}{n} \sum_{k=1}^n X_k \to 0$$

in probability, almost surely, and in quadratic mean.

(b) Examine the convergence of $\bar{X}_n$ in law for $a = 1/2$.

Answer: (a) $E(X_k) = 0$, $V(X_k) = \sigma_k^2 = k^{2a}$,

$$V(\bar{X}_n) = \frac{1}{n^2} \sum_{k=1}^n k^{2a} = O(n^{2a-1}) \to 0.$$

Therefore $2a - 1 < 0$, i.e., $a < 1/2$. This also suffices for

$$\bar{X}_n \xrightarrow{\text{i.m.}} 0.$$

For the Strong Law of Large Numbers (SLLN) it is enough that

$$\sum_{k=1}^n \frac{\sigma_k^2}{k^2} < \infty \implies a < 1/2.$$

(b) Let $\varphi_n(t)$ be the characteristic function of $\bar{X}_n$. We have

$$\log \varphi_n(t) = \sum_{k=1}^n \log \cos k^a \frac{t}{n} = \sum_{k=1}^n \log \left(1 - \frac{k^{2a}t^2}{2n^2} + O\left(\frac{k^{4a}t^4}{n^4}\right)\right)$$

$$= \sum_{k=1}^n \left(-\frac{k^{2a}t^2}{2n^2} + O\left(\frac{k^{4a}t^4}{n^4}\right)\right) = -\frac{n(n+1)}{4n^2}t^2 + O\left(\frac{t^2}{n}\right),$$

because $a = 1/2$. For $|t| \leq \delta$, $\log \varphi_n(t) \to -t^4/4$ uniformly. Therefore

$$\bar{X}_n \to N\left(0, \frac{\sqrt{2}}{2}\right).$$

4.18.* By the so-called *Monte Carlo method*, how many trials are needed to estimate the integral

$$I = \int_0^{\pi/2} \sin x \, dx,$$

so that the absolute error of the esimtate will not exceed 0.1% of I, with probability $p \geq 0.99$?

   Answer: The integral

$$\frac{2}{\pi} I = \frac{2}{\pi} \int_0^{\pi/2} \sin x \, dx,$$

can be considered as the mean value of the sin x, where the random variable X is uniform on $(0, \pi/2)$. So an approximate value of I is

$$I_n = \frac{\pi}{2n} \sum_{k=1}^n \sin X_n,$$

where X_n are (pseudo) random numbers in $(0, \pi/2)$. Now, using

$$\frac{I_n - I}{\sigma(I_n)} \sim N(0, 1),$$

where

$$\sigma^2(I_n) = \frac{\pi^2}{4n} V(\sin X) = \frac{\pi^2 - 8}{8n},$$

we find:

$$n \simeq 1.55 \times 10^6 = 1{,}550{,}000.$$

4.19. Suppose V is a region of a plane with area one and f defined in V with $|f(x, y)| \leq c$. For the calculation of $I = \iint_V f(x, y) \, dx \, dy$, by the Monte Carlo method, we randomly choose n *points* $(x_1, y_1), (x_2, y_2), \ldots, (x_n, y_n)$ in V and we calculate I according to the formula

$$I_n = \frac{1}{n} \sum_{i=1}^n f(x_i, y_i).$$

Prove that

$$E(I_n) = I, \qquad V(I_n) \leq \frac{c^2}{4n}.$$

5. Miscellaneous Complements and Problems

5.1. *Borel–Cantelli Lemma.* In the study of sequences of events $A_1, A_2, \ldots$ with $p_k = P[A_k]$, a significant role is played by the Borel–Cantelli lemma:

(a) If the series $\sum p_k$ converges, then a finite number of the events A_k occurs with probability 1.

(b) If the events are (completely) independent and the series $\sum p_k$ diverges, then an infinite number of the events A_k occur with probability 1.

In the study of the divergence of series of random variables the following theorem is very helpful.

5.2. *Kolmogorov's Three-Series Theorem.* Suppose $\{X_n, n \geq 1\}$ is a sequence of independent random variables, and for any $c > 0$, consider the truncated random variables

$$X'_n = \begin{cases} X_n & \text{if } |X_n| \leq c, \\ 0 & \text{otherwise.} \end{cases}$$

Then the series

$$S = \sum_{n=1}^{\infty} X_n$$

converges with probability 1 if and only if the following *three series* converge:

$$\sum P(|X_n| > c) < \infty, \qquad \sum V(X_n) < \infty \qquad \text{and} \qquad \sum E(X'_n).$$

Otherwise, S converges with probability 0.

The following "laws" are also worth mentioning:

5.3. *The Law of the Iterated Logarithm.* This law concerns the frequency of appearance of large values of the standardized number of successes,

$$S_n^* = \frac{S_n - np}{\sqrt{npq}},$$

in an infinite sequence of Bernoulli trials. According to the De Moivre–Laplace Central Limit Theorem (CLT), we have

$$P[S_n^* > x] \sim 1 - \Phi(x),$$

where $\sim$ means that the ratio of the two sides tends to one.

Therefore, for a particular n, large values (> 4) of S_n^* are improbable, but, obviously, for large n, it is possible for S_n^* to exceed, sooner or later, any large value. How soon this may occur is given by the Law of the Iterated Logarithm (Khintchine; see, e.g., Feller, 1957). With probability 1

$$\limsup_{n \to \infty} \frac{S_n^*}{\sqrt{2 \log \log n}} = 1, \tag{1}$$

i.e., for $\lambda > 1$, only a finite number of the events

$$S_n > np + \lambda\sqrt{npq}\sqrt{2\log\log n} \tag{2}$$

occurs with probability 1, while for $\lambda < 1$, (2) holds for infinitely many n with probability 1.

For reasons of symmetry (of the distribution of S_n^*), (1) implies

$$\liminf_{n\to\infty} \frac{S_n^*}{\sqrt{2\log\log n}} = -1.$$

The behavior of S_n is further illustrated by the following:

(a) There exists a constant $c > 0$ that depends on p, but not on n, so that for all n

$$P[S_n > np] > c.$$

Hint: The probability, according to the binomial distribution, is always positive and, according to the CLT of Laplace, it approaches $1/2$.

(b) Suppose x ($0 \le x < 1$), with decimal expansion

$$x = 0, a_1 a_2 a_3, \ldots, \tag{3}$$

where each a_i is one of the digits $0, 1, \ldots, 9$. Let $a_i = 0$ (instead of 0 we could have chosen any other digit) define a success with $p = 1/10$. Therefore, (3) corresponds to an infinite sequence of Bernoulli trials with $p = 1/10$, and all the limit theorems concerning Bernoulli trials with $p = 1/10$ can be translated into theorems on decimal expansions. $S_n(x)$, the number of zeros among the first n decimal digits of x, takes the place of S_n. Show that

(i) $S_n(x)/x \to 0.1$ in measure (Lebesgue) (in probability).
(ii) $S_n(x)/x \to 0.1$ almost everywhere (with probability 1).
(iii) $\displaystyle\limsup_{n\to\infty} \frac{S_n - n/10}{(n\log\log n)^{1/2}}$.

Answer: (i) WLLN, (ii) SLLN, (iii) Application of (1).

5.4. *The Zero–One Law of Kolmogorov*. Let $\{X_n, n \ge 1\}$ be a sequence of independent random variables and A an event independent of (any event defined in terms of) $X_1, \ldots, X_k$ for every k. Then, either $P(A) = 0$ or $P(A) = 1$.

5.5.* *The tail of the normal $N(0, 1)$.* For large x ($x \to \infty$), the approximation

$$1 - \Phi(x) \sim \frac{1}{x}\varphi(x) \tag{4}$$

is valid; more precisely, for every $x > 0$, the double inequality

$$\varphi(x)\left(\frac{1}{x} - \frac{1}{x^3}\right) < 1 - \Phi(x) < \varphi(x)\frac{1}{x} \tag{5}$$

is valid. Moreover, for every constant $a > 0$ show that, as $x \to \infty$,

$$\left\{1 - \Phi\left(x + \frac{a}{x}\right)\right\} \div \{1 - \Phi(x)\} \to e^{-a}.$$

Hint: Verify that

$$\frac{1}{x}\varphi(x) = \int_x^\infty \varphi(y)(1 + y^{-2})\, dy > \int_x^\infty \varphi(y)\, dy = 1 - \Phi(x)$$

and, similarily, make use of the integral of

$$\varphi(y)(1 - 3y^{-4}).$$

5.6. Consider a sequence $\{A_n, n \geq 1\}$ of independent events. Then a finite or infinite number of events A_k occur with probability 1, according to whether the series $\sum_n P(A_n)$ converges or diverges, respectively (see the above Borel–Cantelli Lemma). Show this by using the Three-Series Theorem.

5.7. Let $\{X_n, n \geq 1\}$ be a sequence of random variables with $\mu_n = E(X_n) < \infty$ and X a random variable with $V(X) < \infty$. If, for each k, $X_1, \ldots, X_k$ and $X - (X_1 + \cdots + X_k)$ are independent, prove that $V(X_k) < \infty$ for every k, and that the series

$$\sum (X_k - E(X_k))$$

converges with probability 1 (almost everywhere or almost surely).

5.8. In a sequence of (independent) Bernoulli trials with probability of success p, the event A_k is realized if k consecutive successes appear between the 2^kth and 2^{k+1}th trial. Prove that if $p \geq 1/2$, then an infinite number of the events A_k occur with probability 1, whereas for $p < 1/2$ with probability 1 a finite number of the events A_k occur.

5.9.* *Bochner–Khintchine Theorem.* A continuous function $\varphi(t)$ with $\varphi(0) = 1$ is a characteristic function if and only if it is nonnegative definite, i.e., if for each $n \geq 1$ and for each n-tuple of real numbers $t_1, \ldots, t_n$ and complex numbers $Z_1, \ldots, Z_n$ it satisfies the

$$\sum_{j=1}^{v} \sum_{k=1}^{v} \varphi(t_j - t_k)Z_j\bar{Z}_k \geq 0. \tag{*}$$

Prove that $(*)$ is necessary.
    Hint: The left side of $(*)$ is equal to

$$\int \left(\sum_k e^{it_k x}Z_k\right)\left(\sum_j e^{-it_j x}Z_j\right) dF(x),$$

where F is some distribution function. For a proof of the sufficiency of $(*)$ see, e.g., Gnedenko (1962).

5.10. Prove that the functions, (a) $e^{-i|t|}$, (b) $1/(1 - i|t|)$, (c) $\cos t^2$, are not characteristic functions.

5.11. If $\varphi(t)$ is a characteristic function, prove that $g(t) = e^{\varphi(t)-1}$ is also a characteristic function.

Hint: See (6.6).

5.12. Show that, for a real characteristic function $\varphi(t)$, the following inequalities are valid:

(a)
$$1 - \varphi(nt) \leq n^2(1 - \varphi(t)), \qquad n = 0, 1, 2, \ldots$$

(b)
$$1 + \varphi(2t) \geq 2[\varphi(t)]^2.$$

5.13. Let X be the number of Bernoulli trials required until we have observed r *successive* successes. Find the generating function of X and $E(X)$.

Answer:
$$P(t) = \sum_{n=r}^{\infty} t^n P[X = n] = \sum_{n=r}^{\infty} t^n p_{n,r} = \frac{p^r t^r (1 - pt)}{1 - t + p^r q t^{r+1}},$$

$$E(X) = P'(1) = \frac{1 - p^r}{p^r q^r}.$$

5.14.* Let M_n be the maximum number of consecutive successes (the maximum *length of a run*) observed in n Bernoulli trials. If

$$P_{n,r} = P[M_n \leq r],$$

show that

$$P_{n,r} = 1 - p_{1,r} - p_{2,r} - \cdots - p_{n,r},$$

where $p_{n,r}$ was defined in the preceding problem, and therefore, the generating function of $p_{n,r}$ is

$$\sum_{n=r}^{\infty} P_{n,r} t^n = \frac{1 - P(t)}{1 - t} = \frac{1 - p^r t^r}{1 - t + p^r q t^{r+1}}.$$

Also, show that

$$E(M) = \frac{\log n}{-\log p} + O(1),$$

$$V(M) = O(1).$$

5.15. Find the distributions with characteristic functions, (a) $\cos t$, (b) $\cos^2 t$, (c) $\sum_{k=0}^{\infty} p_k \cos kt$ where $p_k \geq 0$, $\sum p_k = 1$.

Answer: (a) $\cos t = \frac{1}{2}[e^{it(-1)} + e^{it(1)}]$. Therefore

$$F(x) = \begin{cases} 0 & \text{if } x < -1, \\ 1/2 & \text{if } -1 \leq x < 1, \\ 1 & \text{if } x \geq 1. \end{cases}$$

(b) $\cos^2 t = \frac{1}{4}e^{it(-2)} + \frac{1}{2} + \frac{1}{2}e^{it2}$

$$F(x) = \begin{cases} 0 & \text{if } x < -2, \\ 1/4 & \text{if } -2 \le x < 0, \\ 3/4 & \text{if } 0 \le x < 2, \\ 1 & \text{if } x \ge 2. \end{cases}$$

(c) $\sum p_k \cos kt = \frac{1}{2}\sum p_k e^{-ikt} + \frac{1}{2}\sum p_k e^{ikt}$.
Therefore this is a discrete distribution with jumps $\frac{1}{2}p_k$ at the points $\pm k$ $(k = 0, 1, \ldots)$.

5.16. Find the discrete distributions with generating functions, (a) $\frac{1}{8}(1 + s)^3$, (b) $\frac{1}{2}(1 - \frac{1}{2}s)^{-1}$, (c) $e^{(s-1)}$, (d) $(\frac{1}{4}s + \frac{3}{4})^{100}$.
Answer: (a) Discrete with probabilities 1/8, 3/8, 3/8, 1/8 at the points 0, 1, 2, 3, respectively.
   (b) The probability of the value k is 2^{-k-1} $(k = 0, 1, 2, \ldots)$.
   (c) Poisson with $\lambda = 1$.
   (d) Binomial with $p = 1/4$ and $n = 100$.

5.17. Using the Kolmogorov inequality (8.9), show that if the series

$$\sum_{k=1}^{\infty} V(X_k)/k^2 < \infty,$$

then the SLLN holds, i.e.,

$$\frac{1}{n}\sum_{k=1}^{n} [X_k - E(X_k)] \xrightarrow{\text{w.p.}1} 0.$$

5.18. *A fair but loss-incurring game (a probability "paradox").* The probability that, in each play of a game, the player receives 2^k dollars is

$$p_k = \frac{1}{2^k k(k + 1)}, \qquad k = 1, 2, \ldots, \tag{1}$$

and that he receives 0 dollars is $p_0 = 1 - (p_1 + p_2 + \cdots)$. So, the expected (average) gain in each game is

$$\mu = \sum_{k=1}^{\infty} 2^k p_k = \sum_{k=1}^{\infty} \frac{1}{k(k + 1)} = \left(1 - \frac{1}{2}\right) + \left(\frac{1}{2} - \frac{1}{3}\right) + \left(\frac{1}{3} - \frac{1}{4}\right) + \cdots = 1.$$

If, in each game, the player pays a 1-dollar fee, the net profit of the player after n games is equal to

$$\sum_{k=1}^{n} X_k - n = S_n - n, \qquad \text{with} \quad E(S_n - n) = 0,$$

i.e., the game is fair (X_k is a random variable with distribution (1)). For every $\varepsilon > 0$, however, the probability that, in n games, the player will lose more than $(1 - \varepsilon)n/\log_2 n$ dollars, approaches 1, i.e., it can be proved (truncating the random variables X_k, see Feller (1957) or Gnedenko (1962)) that

$$\lim_{n \to \infty} P\left[S_n - n < \frac{(1 - \varepsilon)n}{\log_2 n} \right] = 1.$$

5.19. In a sequence $\{X_n, n \geq 1\}$ of Bernoulli random variables, suppose that

$$Y_n = 0 \quad \text{if } X_n X_{n+1} = 1 \text{ or } X_n = 0 \text{ and } X_{n+1} = 0,$$
$$Y_n = 1 \quad \text{if } X_n = 1 \text{ or } X_{n+1} = 1.$$

Find $E(Z_n)$ and $V(Z_n)$ where $Z_n = \sum_{i=1}^{n} Y_i$.
   Answer:

$$E(Z_n) = nE(Y_1) = 2npq,$$
$$V(Z_n) = 2npq(1 - 2pq) + 2(n - 1)pq(p - q)^2.$$

5.20. We randomly choose two numbers in $(0, 1)$. What is the probability p that their sum be smaller than 1 and their product smaller than 3/16?
   Answer: $p = $ area S, where $S = \{(x, y): x \geq 0, y \geq 0, x + y \leq 1, xy \leq 3/16\}$.

$$p = \frac{1}{4} + \frac{3}{16} \log 3 = 0.456.$$

5.21. A bus on line A arrives at a bus station every 4 minutes and a bus on line B every 6 minutes. The time interval between an arrival of a bus for line A and a bus for line B is uniformly distributed between 0 and 4 minutes. Find the probability:
   (a) that the first bus that arrives will be for line A;
   (b) that a bus will arrive within 2 minutes (for line A or B).
   Answer: Suppose x is an instant of time, where $0 \leq x \leq 12$ minutes. The times of arrival of the buses for line A are $x = 0, 4, 8$, and of buses for line B the arrival times are $y, y + 6$ with $0 \leq y \leq 4$.
   (a) The favorable cases are: for $0 < y \leq 2$ we have $y < x \leq 4$ or $6 + y \leq x \leq 12$; and for $y > 2$ we have $y < x < 8$ or $y + 6 < x < 12$. Therefore $p = 2/3$.
   (b) Favorable cases: $2 \leq x \leq 4, 6 \leq x \leq 8, 10 \leq x \leq 12, 4 + y \leq x \leq 6 + y$. For $y < 2$ we have $0 < x < y$, and for $y > 2$ we have $y - 2 \leq x \leq y$. Therefore $p = 2/3$.

5.22. N stars are randomly scattered, independently of each other, in a sphere of radius R.
   (a) What is the probability that the star nearest to the center is at a distance at least r?
   (b) Find the limit of the probability in (a) if

$$R \to \infty \qquad \text{and} \qquad N/R^3 \to 4\pi\lambda/3.$$

   Answer: (a) $(1 - r^3/R^3)^N$.
   (b) $\exp\{-4\pi\lambda r^3/3\}$ (close to the sun $\lambda \approx 0.0063$, when R is measured in parsecs).

5.23. A satellite moving on an orbit between two parallels 60° north and 60° south (latitude) is equally likely to land at any point between the two parallels. What is the probability p that the satellite will land in the north above 30°?

Answer:

$$p = \left\{ R^2 \int_0^{2\pi} d\theta \int_{\pi/6}^{\pi/3} \cos\varphi \, d\varphi \right\} \bigg/ \left\{ 2R^2 \int_0^{2\pi} d\theta \int_0^{\pi/3} \cos\varphi \, d\varphi \right\} = 0.21.$$

5.24. In the equation $\lambda^3 + 3X\lambda + Y = 0$, the coefficients X, Y are uniformly distributed in the rectangle $|X| \le a$, $|Y| \le b$. What is the probability p that the equation has real roots?

Answer: It is required that $Y^2 + X^3 \le 0$, i.e., for $X \le 0$, when $Y^2 \le -X^3$. If $a^3 \le b^2$, we have

$$p = \frac{1}{2ab} \int_0^a x^{3/2} \, dx = \frac{a^{3/2}}{5b}.$$

If $a^3 \ge b^2$, we have

$$p = \frac{1}{2} - \frac{1}{2ab} \int_0^b y^{2/3} \, dy = \frac{1}{2}\left(1 - \frac{3b^{2/3}}{a}\right).$$

5.25.* *The binomial distribution via difference equations.* Let $P_{k,n}$ denote the probability of k successes in n independent Bernoulli trials. By using the generating function $G_n(t)$ of $P_{n,k}$ ($k = 0, 1, 2, \ldots$), deduce the binomial distribution.

Hint: $P_{k,n}$ satisfies the partial-difference equation

$$P_{k,n} = pP_{k-1,n-1} + qP_{k,n-1}, \qquad P_{k,0} = 0, \quad k \ne 0, \quad P_{0,n} = q^n,$$

from which we can find the difference equation for G_n:

$$G_n(t) = (pt + q)G_{n-1}(t), \qquad G_0(t) = 1,$$

which has the solution

$$G_n(t) = (pt + q)^n,$$

i.e., the probability generating function of the binomial.

5.26. A and B play the following game. They flip a coin, if the outcome is a head A gets 1 dollar from B, otherwise A pays 1 dollar to B. Initially, each of them has 3 dollars. The game terminates when either A or B loses all his money. What is the probability p_n that n flips are needed?

Answer: $p_n > 0$ only for $n = 2k + 1$ ($k = 1, 2, \ldots$). Suppose q_k is the probability that the game will not finish after $2k + 1$ flips. Then

$$p_n = \frac{1}{4}p_{(n-3)/2} \quad \text{with} \quad p_k = \left(\frac{3}{4}\right)^k, \qquad \text{i.e.,} \quad p_n = \frac{1}{4}\left(\frac{3}{4}\right)^{(n-3)/2}.$$

5.27. Let $Y_n = \max\{X_1, \ldots, X_n\}$ where $X_1, \ldots, X_n$ are independent and identically distributed random variables with a uniform distribution on $(0, 1)$.

Show that the distribution of

$$Z_n = n(1 - Y_n)$$

converges (as $n \to \infty$) to the exponential distribution with distribution function $F(z) = 1 - e^{-z}$.

Answer: Examine the limit of the sequence $\{F_n(z)\}$ of the distribution functions of $\{Z_n\}$.

5.28. A discrete random variable X_n that appears in the theory of extremes has the distribution function

$$F_n(n) = 1 - \frac{(n)_r}{(n + nx)_r}, \qquad 1 \le r \le n,$$

and X_n takes the values $1/n, 2/n, \ldots$. Show that the sequence $\{X_n\}$ converges in law to a continuous distribution with distribution function

$$F(x) = \begin{cases} 0, & x \le 0, \\ 1 - (1 + x)^{-r}, & x > 0. \end{cases}$$

5.29. If $X_n \overset{P}{\to} X$ and $E(X_n - Y_n)^2 \to 0$, show that the sequence $\{Y_n\}$ also converges to X in probability.

PART III
SOLUTIONS

Solutions

1. $(A_1 \cup A_2)F$:   The set of female students in the freshman and sophomore years.

$C'F = F - C$:   The set of non-Cypriot female students.

$A_1F'C = A_1C - F$: The set of Cypriot male freshmen students.

$A_3FC' = A_3F - C$: The set of non-Cypriot junior female students.

$(A_1 \cup A_2)CF$:   The set of Cypriot female students in the freshman and sophomore years.

2. (a) $(A \cup B)(A \cup C) = A \cup (BC)$.

(b) $(A \cup B)(A' \cup B) = B \cup (AA') = B \cup \emptyset = B$.

(c) $(A \cup B)(A' \cup B)(A' \cup B') = B(A' \cup B') = B(AB)' = B - A$ (by de Morgan's laws and (b)).

3.

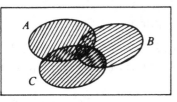

(a) The shaded area represents $A \cup B \cup C$.

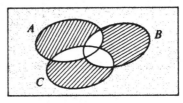

(b) The shaded area represents $(AB'C') \cup (A'BC') \cup (A'B'C) \cup (A'B'C')$.

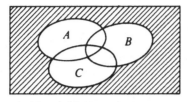

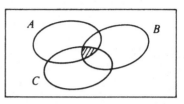

(c) The shaded area represents $(A \cup B \cup C)' = A'B'C'$.

(d) The shaded area represents ABC.

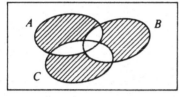

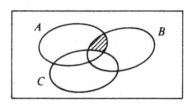

(e) The shaded area represents $(AB'C') \cup (A'BC') \cup (A'B'C)$.

(f) The shaded area represents $ABC' = AB - C$.

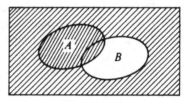

(g) The shaded area represents $A \cup A'B'$.

4. (a) $A = \{(HHT), (THH), (HTH)\}$,

$B = \{(HHT), (THH), (HTH), (HHH)\}$,

$C = \{(THH), (THT), (TTH)\}$.

(b) (i) $A'B = \{(HHH)\}$: exactly three heads.

(ii) $A'B' = (A \cup B)' = B' = \{(HTT), (TTT). (TTH), (THT)\}$: at least two tails.

(iii) $AC = \{(THH)\}$. The first is tails, and the second and third are heads.

5. Denoting by M and F a male and a female child, respectively, we have

$A = \{MFMF, FMFM\}$,

$B = \{MMMM, MMFM, MFMM, MFFM\}$,

$C = \{MMFF, MFFM, MFMF, FFMM, FMFM, FMMF\}$,

$D = \{MMMM, MMMF, FMMM, FFFF, FFFM, MFFF\}$.

6. $A = \{abcbca, acbcba\}$.

7. 4^4.

$A = \{aaaa, bbbb, cccc, dddd\}$;

$B = \{aabc, aacb, abbc, \ldots\}$,   that is, $\binom{4}{2}\binom{2}{1}\binom{4}{1}\binom{3}{2} = 144$ points,

where a, b, c, and d denote the four types of cellular disorders, and the position of the letter denotes the patient.

8. 3^2. $A = \{ab*, ba*, a*b, b*a, *ab, *ba\}$,
                    $B = \{ab*, a*b, ba*, b*a\}$,

where a, b denote the two persons, $*$ means "none", and the position of a, b, $*$ indicates the floor, e.g., $a*b$ means: a at first floow and b at third floor.

9. $\binom{10}{1}\binom{5}{1}\binom{2}{1} = 100$ (by the multiplication rule).

10. $4! \cdot (3!)^2 (4!)^2$ (i.e., permutations of the four works times permutations of volumes within each of the four works.

11. Identify the n persons with n cells and the r objects with r balls to be placed in the n cells:
   (i) if the objects are distinguishable there are n^r ways;
   (ii) if the objects are indistinguishable there are $\binom{n+r-1}{r}$ ways.

12. The 5 boys define consecutive gaps (positions) between any two of them; the 5 girls can be placed in the positions in as many ways as the number of permutations of 5 things taken all at a time, i.e., $5! = 120$.
   The 5 boys can be seated at a round table in as many ways as the number of cyclical permutations of 5, i.e., $(5 - 1)! = 4! = 24$. Therefore, the total number of ways is equal to $4! \, 5! = 2,880$.

13. As many chords as the number of combinations $\binom{8}{2}$, i.e., 28. As many triangles as the number of combinations $\binom{8}{3}$, i.e., 56. As many hexagons as $\binom{8}{6} = 28$.

14. Cyclical permutations of n, i.e., $(n - 1)!$. The probability that any fixed person sits right or left of any given person is the same for every person; and hence is equal to $1/(n - 1)$. Therefore, the probability sought is $2/(n - 1)$. This also follows from the fact that out of the $(n - 1)!$ total cases there are $2(n - 2)!$ favorable cases.

15. By virtue of (1.5) the 20 recruits can be distributed into 4 camps in $20!$ $(5!)^4$ ways. However, to each division in 4 groups, there correspond $4!$ distributions to 4 camps. Hence, there are

$$\frac{20!}{(5!)^4 4!}$$

distributions into 4 groups of 5 each.

16. According to the multiplication formula, we can form

$$\binom{7}{4} \cdot \binom{5}{3} \cdot 7!$$

words (every permutation of the seven letters is supposed to be a word).

17. (i) Pascal triangle. (ii) Use (i) repeatedly.

18. The n things are divided into two groups of m and $n - m$. Any r of the n will contain k of the m and $r - k$ of the $n - m$ where $k = 0, 1, 2, \ldots, r$. Hence the expression for $\binom{n}{r}$.

19. (i) The expansion of $(1 - 1)^n = 0$.
   (ii) Differentiate $(1 + x)^n$ once and put $x = 1$.
   (iii) Differentiate $(1 + x)^n$ twice and put $x = 1$.

20. (i) Expand the right-hand side. (ii) Use 17(i).

21.
$$\binom{2n}{n} = \sum_{k=0}^{n} \binom{n}{k}\binom{n}{n-k} = \sum_{k=0}^{n} \binom{n}{k}^2.$$

22.
$$\sum_{k=0}^{n} \frac{(2n)!}{(k!)^2(n-k)!^2} = \sum_{k=0}^{n} \binom{2n}{n}\binom{n}{k}^2 = \binom{2n}{n}\sum_{k=0}^{n} \binom{n}{k}^2 = \binom{2n}{n}^2.$$

25. Examine the inequality $a_{k+1}/a_k \geq 1$. It gives $k \leq (nx - 1)/(x + 1)$ and the required maximum value of a_k is the a_{k^*+1} with

$$k^* = \left[\frac{nx - 1}{x + 1}\right],$$

where $[a]$ denotes the integer part of a. If $(nx - 1)/(x + 1)$ is an integer, r say, then there are two maxima, namely, the a_r and a_{r+1}. On the other hand, if $x < 1/n$ then $a_{k+1} < a_k$ for every k and max $a_k = a_0 = 1$. For $e = 2.72, n = 100$ we have

$$k^* = \left[\frac{100e - 1}{e + 1}\right].$$

26.
$$P(A') = 1 - P(A) = 1 - \tfrac{1}{3} = \tfrac{2}{3},$$

$$P(A' \cup B) = P(A') + P(B) - P(A'B)$$

$$= 1 - P(A) + P(B) - \{P(B) - P(AB)\}$$

$$= 1 - P(A) + P(AB) = 1 - \tfrac{1}{3} + \tfrac{1}{6} = \tfrac{5}{6},$$

$$P(A \cup B') = P(A) + P(B') - P(AB')$$
$$= P(A) + 1 - P(B) - \{P(A) - P(AB)\}$$
$$= 1 - P(B) + P(AB) = 1 - \tfrac{1}{4} + \tfrac{1}{6} = \tfrac{11}{12},$$
$$P(A'B') = P[(A \cup B)'] = 1 - P(A \cup B)$$
$$= 1 - P(A) - P(B) + P(AB) = 1 - \tfrac{1}{3} - \tfrac{1}{4} + \tfrac{1}{6} = \tfrac{7}{12},$$
$$P(A' \cup B') = P[(AB)'] = 1 - P(AB) = 1 - \tfrac{1}{6} = \tfrac{5}{6}.$$

27. From $A \subseteq A \cup B$, $B \subseteq A \cup B$, we conclude that $P(A) \leq P(A \cup B)$, $P(B) \leq P(A \cup B)$. Hence

$$P(A \cup B) \geq \max\{P(A), P(B)\}, \tag{1}$$

and as $P(A) = 3/4 > P(B) = 3/8$ it follows that $P(A \cup B) \geq 3/4$. Because of $A \supseteq AB$, $B \supseteq AB$ we have $P(AB) \leq P(A)$, $P(AB) \leq P(B)$ and so

$$P(AB) \leq \min(P(A), P(B)). \tag{2}$$

Consequently, $P(AB) \leq 3/8$.

By the addition theorem we have $P(AB) + P(A \cup B) = P(A) + P(B)$ and, because of $P(A \cup B) \leq 1$, $P(AB) \geq P(A) + P(B) - 1$, but $P(AB) \geq 0$ and hence

$$P(AB) \geq \max\{0, P(A) + P(B) - 1\}.$$
$$P(AB) \geq \tfrac{3}{4} + \tfrac{3}{8} - 1 = \tfrac{1}{8}. \tag{3}$$

From (1), (2), and (3) for $P(A) = 1/3$, $P(B) = 1/4$ we have the inequalities

$$P(A \cup B) \geq \tfrac{1}{3} \quad \text{and} \quad 0 \leq P(AB) \leq \tfrac{1}{4}.$$

28. We have $P(A'B) = P(B) - P(AB)$ and $P(A) = 1 - P(A')$. Hence

$$P(AB) - P(A)P(B) = P(AB) - [1 - P(A')]P(B)$$
$$= P(A')P(B) - [P(B) - P(AB)]$$
$$= P(A')P(B) - P(A'B). \tag{1}$$
$$P(AB) - P(A)P(B) = P(AB) - P(A)[1 - P(B')]$$
$$= P(A)P(B') - [P(A) - P(AB)]$$
$$= P(A)P(B') - P(AB'). \tag{2}$$

The required relation follows from (1) and (2).

29. By the addition theorem, we have

$$P(A_1 A_2) = P(A_1) + P(A_2) - P(A_1 \cup A_2) \geq P(A_1) + P(A_2) - 1. \tag{1}$$

That is, for $n = 2$ the inequality holds. Suppose that it holds for $n = k$, i.e.,

$$P(A_1 A_2 \ldots A_k) \geq \sum_{i=1}^{k} P(A_i) - (k - 1).$$

It is sufficient to show that it also holds for $n = k + 1$. Using (1) and (2) we get

$$P(A_1 A_2 \ldots A_{k+1}) = P[(A_1 A_2 \ldots A_k)A_{k+1}] \le P(A_1 A_2 \ldots A_k) + P(A_{k+1}) - 1$$

$$\ge \sum_{i=1}^{k} P(A_1) - (k - 1) + P(A_{k+1}) - 1 = \sum_{i=1}^{k+1} P(A_i) - k.$$

30. We have

$$P[A \cup B \cup C] = P[A \cup (B \cup C)] \qquad \text{and by the addition theorem}$$

$$= P(A) + P(B \cup C) - P[A(B \cup C)]$$

$$= P(A) + P(B \cup C) - P[(AB) \cup (AC)]$$

$$= P(A) + P(B) + P(C) - P(BC) - P(AB) - P(AC)$$

$$+ P(ABC).$$

$P[\text{at least one}] = P(A \cup B \cup C) = 0.5 + 0.4 + 0.3 - 0.35 - 0.25 - 0.2 + 0.15 = 0.55$, that is, 55%.

31. (a) Let $p_n(k)$ denote the probability that the integer n is divisible by k. Then $p_n(k) = (1/n)[n/k]$ where $[x]$ denotes the integer part of x. Hence the required probability equals

$$\frac{1}{n}\left(\left[\frac{n}{3}\right] + \left[\frac{n}{4}\right] - \left[\frac{n}{12}\right]\right). \qquad (*)$$

(b) For $n = 100$ we get from $(*)$

$$\frac{1}{100}\left(\left[\frac{100}{3}\right] + \left[\frac{100}{4}\right] - \left[\frac{100}{12}\right]\right) = \frac{1}{100}(33 + 25 - 8) = \frac{1}{2}.$$

(c) Taking the limits as $n \to \infty$ in $(*)$ we obtain

$$\frac{1}{3} + \frac{1}{4} - \frac{1}{12} = \frac{1}{2}.$$

32. (i) Let $p_k = P[\text{exactly } k \text{ of the events } A, B, C \text{ occur}]$ $(k = 0, 1, 2, 3)$, then

$$p_0 = P[A'B'C'] = P[(A \cup B \cup C)'] = 1 - P(A \cup B \cup C) = 1 - P(A) - P(B)$$

$$- P(C) + P(AB) + P(AC) + P(BC) - P(ABC),$$

$$p_1 = P[AB'C' \cup A'BC' \cup A'B'C] = P[A(B \cup C)'] + P[B(C \cup A)']$$

$$+ P[C(A \cup B)']$$

$$= P(A) - P[AB \cup AC] + P(B) - P(BC \cup BA) + P(C) - P[CA \cup CB]$$

$$= P(A) - P(AB) - P(AC) + P(ABC) + P(B) - P(BC) - P(BA)$$

$$+ P(ABC) + P(C) - P(CA) - P(CB) + P(ABC)$$

$$= [P(A) + P(B) + P(C)] - 2[P(AB) + P(AC) + P(BC)] + 3P(ABC),$$

$$p_2 = P[ABC' \cup AB'C \cup A'BC] = P(ABC') + P(ACB') + P[BCA']$$

$$= P(AB) - P(ABC) + P(AC) - P(ABC) + P(BC) - P(ABC)$$

$$= [P(AB) + P(AC) + P(BC)] - 3P(ABC),$$

$$p_3 = P(ABC).$$

(ii) $P[\text{at least } k \text{ of the events } A, B, C \text{ occur}] = \sum_{n=k}^{3} p_n$ the p_i as in (i); see also Exercise 30.

33. By the addition theorem, for $n = 2$, (1.1) holds. Suppose that it holds for $n = r$. We shall show that it also holds for $n = r + 1$. By the addition theorem we have

$$P\left[\bigcup_{i=1}^{r+1} A_i\right] = P\left[\left(\bigcup_{i=1}^{r} A_i\right) \cup A_{r+1}\right]$$

$$= P\left[\bigcup_{i=1}^{r} A_i\right] + P(A_{r+1}) - P\left[\bigcup_{i=1}^{r} (A_i A_{r+1})\right]. \tag{1}$$

By hypothesis for $n = r$ it follows that

$$P\left[\bigcup_{i=1}^{r} (A_i A_{r+1})\right] = \sum_{k=1}^{r} (-1)^{k-1} S_k^*$$

where $\qquad S_k^* = \sum_{1 \le i_1 < i_2 < \cdots < i_k \le r} P(A_{i_1} A_{i_2} \dots A_{i_k} A_{r+1}).$

Let

$$S_k^{**} = \sum_{1 \le i_1 < \cdots < i_k \le r+1} P(A_{i_1} A_{i_2} \dots A_{i_k}), \qquad k = 1, \dots, r+1. \tag{2}$$

Then

$$S_1 + P(A_{r+1}) = \sum_{i=1}^{r} P(A_i) + P(A_{r+1}) = \sum_{i=1}^{r+1} P(A_i) = S_1^{**}, \tag{3}$$

and, in general,

$$S_k + S_{k-1}^* = \sum_{1 \le i_1 < \cdots < i_k \le r} P(A_1 A_2 \dots A_{i_k})$$

$$+ \sum_{1 \le i_1 < \cdots < i_{k-1} \le r} P(A_{i_1} A_{i_2} \dots A_{i_{k-1}} A_{r+1})$$

$$= S_k^{**}, \qquad k = 2, \dots, r, \tag{4}$$

$$S_r^* = P(A_1 A_2 \dots A_r A_{r+1}) = S_{r+1}^{**}. \tag{5}$$

By virtue of (2)–(5), (1) becomes

$$P\left(\bigcup_{i=1}^{r+1} A_i\right) = \sum_{k=1}^{r} (-1)^{k-1} S_k + P(A_{r+1}) - \sum_{k=1}^{r} (-1)^{k-1} S_k^*$$

$$= S_1 + P(A_{r+1}) + \sum_{k=2}^{r} (-1)^{k-1}[S_k + S_{k-1}^*] + (-1)^r S_r^*$$

$$= \sum_{k=1}^{r+1} (-1)^{k-1} S_k^{**},$$

i.e., (1) holds for $n = r + 1$.

34. For any two events A and B we have:

(i) $d(A, B) = P(A \triangle B) = P[AB' \cup BA'] \geq 0$ and $d(A, B) = 0$ if $A = B$, that is, d is a nonnegative function.

(ii) $d(A, B) = P(A \triangle B) = P(B \triangle A) = d(B, A)$ (symmetric property).

To complete the proof we have to show that for any events A, B, and C the triangle inequality holds, i.e.,

(iii) $d(A, C) \leq d(A, B) + d(B, C)$,

$$d(A, B) + d(B, C) - d(A, C) = P(AB' \cup BA') + P(BC' \cup CB')$$

$$- P(AC' \cup CA')$$

$$= P(AB') + P(BA') + P(BC') + P(CB')$$

$$- P(AC') - P(CA')$$

$$= P(AB'C \cup AB'C') + P(A'BC \cup A'BC')$$

$$+ P(ABC' \cup A'BC') + P(AB'C \cup A'B'C)$$

$$- P(ABC' \cup AB'C') - P(A'BC \cup A'B'C)$$

$$= P(AB'C) + P(AB'C') + P(A'BC) + P(A'BC')$$

$$+ P(ABC') + P(A'BC') + P(AB'C)$$

$$+ P(A'B'C) - P(ABC') - P(AB'C')$$

$$- P(A'BC) - P(A'B'C)$$

$$= 2[P(AB'C) + P(A'BC')]$$

$$\geq 0.$$

35. (a) The probability p_0 of no lucky ticket is

$$p_0 = \frac{\binom{96}{3}}{\binom{100}{3}}$$

and the probability of winning is $1 - p_0$.

(b) The probability of winning is

$$\frac{\binom{99}{2}}{\binom{100}{3}}.$$

36. $P[\text{an underweight loaf is discovered}] = 1 - \dfrac{\binom{10}{0}\binom{70}{5}}{\binom{80}{5}}.$

37. Let X_j denote the number of persons born on the jth day of the week (among the 7 persons). Then $(X_1, \ldots, X_7)$ has the multinomial distribution

$$P[X_j = n_j, j = 1, 2, \ldots, 7] = \frac{7!}{n_1! \, n_2! \ldots n_7!}\left(\frac{1}{7}\right)^7, \qquad \sum_{j=1}^{7} n_j = 7.$$

(a) $P[X_j = 1, j = 1, 2, \ldots, 7] = 7! \, (1/7)^7.$
(b) $P[\text{at least two on the same day}] = 1 - P[X_j = 1, j = 1, 2, \ldots, 7] = 1 - 7! \, (1/7)^7.$
(c) $P[\text{two on Sunday and two on Tuesday}] = \dfrac{7!}{2! \, 2! \, 3!}\left(\dfrac{1}{7}\right)^4\left(\dfrac{5}{7}\right)^3.$

38. $P = \dfrac{\binom{2N}{N}\binom{2N}{N}}{\binom{4N}{2N}}.$

39. For real roots we must have

$$ac \le \frac{b^2}{4}. \tag{1}$$

From the values of the product ac we observe that inequality (1) holds as follows:

Value of b:	2	3	4	5	6
Favorable cases:	1	3	8	14	17

Thus we have 43 favorable cases out of a total of $6^3 = 216$. Hence the probability of real roots is 43/216 and of complex roots is 173/216.

40. (a) $\dfrac{4}{\binom{52}{5}}$ (a favorable case for each of the 4 suits).

(b) $\dfrac{13 \times 48}{\binom{52}{5}}$ (to each choice of 4 cards of the same face value (ace, 2, 3, etc.) there correspond 48 ways of choosing the fifth card).

(c) $\dfrac{5 \times 4^5}{\binom{52}{5}}$ (there are 5 quintuplets with successive face values, e.g., (2, 3, 4, 5, 6), (3, 4, 5, 6, 7), etc., and each of the 5 cards can be chosen in 4 ways corresponding to the 4 suits).

41. (i) Multiple hypergeometric:

$$\binom{13}{v_1}\binom{13}{v_2}\binom{13}{v_3}\binom{13}{13-v_1-v_2-v_3}\Big/\binom{52}{13}.$$

(ii)
$$\binom{4}{v}\binom{48}{13-v}\Big/\binom{52}{13}, \qquad v = 0, 1, 2, 3, 4.$$

(iii)
$$\binom{4}{v_1}\binom{4}{v_2}\binom{52-8}{13-v_1-v_2}\Big/\binom{52}{13}.$$

(a) Let the event A_j be the jth player who gets an ace ($j = 1, 2, 3, 4$). Then

$$P[A_1 A_2 A_3 A_4] = \frac{4!\,48!/(12!)^4}{52!/(13!)^4}.$$

(b) Let B_j be the event when the jth player gets all the aces ($j = 1, 2, 3, 4$). Then

$$P[\text{a player gets all the aces}] = 4P[B_j] = 4\binom{48}{9}\Big/\binom{52}{13}.$$

(c) $P[\text{a player gets } v_1 \text{ aces and his partner } v_2 \text{ aces}]$

$$= 2 \times \frac{\binom{4}{v_1}\binom{48}{13-v_1}\binom{4-v_1}{v_2}\binom{35+v_1}{13-v_2}}{\binom{52}{13}\binom{39}{13}}$$

(i.e., of the two hands one has v_1 aces and the other has v_2 aces), if $v_1 \neq v_2$, when $v_1 = v_2$ the factor 2 above should be deleted.

42. (a) The ten boys define ten successive intervals (arcs). The six girls can choose six of the intervals in $\binom{10}{6}$ ways. The event A that a certain boy remains between two girls can be realized in $\binom{8}{6}$ ways. Hence

$$P[A] = \binom{8}{6}\Big/\binom{10}{6}.$$

(b) The six girls can enter in $(10)_6$ ways. The event B that certain girl stands by certain boy can be realized in $2 \times (9)_5$ ways. Hence

$$P[B] = \frac{2 \times (9)_5}{(10)_6} = \frac{1}{5}.$$

This is immediate since the girl can choose 2 favorable positions out of the 10. We cannot conclude that her choice is not random.

43. (a) (i) $5pq^{25}$ with $p = 1 - q = 1/26$.

   (ii) $(3/13)^5$.

   (iii) $1/26^5$.

(b) (i) $\dbinom{25}{4} \Big/ \dbinom{26}{5} = \dfrac{5}{26}$.

   (ii) $\dbinom{6}{5} \Big/ \dbinom{26}{5}$.

   (iii) $1/(26)_5$.

44. Let us represent the r balls by r stars, and the n cells by n spaces between $n + 1$ bars. Each distribution starts and ends with a bar and between these extreme bars there are r stars and $n - 1$ bars. To each selection, either of the r places for the stars or of the $n - 1$ places for the bars, there corresponds a distribution. Consequently, there are

$$\binom{n + r - 1}{r} = \binom{n + r - 1}{n - 1} \tag{*}$$

such distributions (Feller, 1957, p. 36). An alternative proof is by induction on n as follows: Suppose $(*)$ holds for $n = k$, i.e., the number of solutions $x_i \geq 0$, x_i integers, of the equation

$$x_1 + \cdots + x_k = r$$

is given by

$$s(k, r) = \binom{k + r - 1}{k - 1}.$$

Then clearly the solutions of

$$x_1 + \cdots + x_{k+1} = r,$$

or equivalently of

$$x_1 + \cdots + x_k = r - x_{k+1},$$

are the solutions of

$$x_1 + \cdots + x_k = r - i, \qquad i = 0, 1, \ldots, r,$$

that is,

$$s(k + 1, r) = s(k, r) + s(k, r - 1) + \cdots + s(k, 0)$$

$$= (k + r - 1) + \binom{k + r - 2}{k - 1} + \cdots + \binom{k - 1}{k - 1},$$

which, by Exercise 17(ii), gives

$$s(k + 1, r) = \binom{k + r}{k},$$

i.e., $(*)$ holds for $n = k + 1$.

45. The number of distributions with exactly m empty cells is $\binom{n}{m} \cdot$ $\binom{r-1}{n-m-1}$. Indeed, if we represent the r balls with r stars and the n cells by spaces between $n+1$ bars, then we can select m cells (empty cells) from the n in $\binom{n}{m}$ ways. Without any loss of generality, let the empty cells be the first m, that is, the spaces between the first $m+1$ bars. The r stars leave $r-1$ spaces of which $n-m-1$ are to be occupied by bars and hence we have $\binom{r-1}{n-m-1}$ choices. Using the result of Exercise 44, we conclude that the required probability equals

$$p_m = \frac{\binom{n}{m} \cdot \binom{r-1}{n-m-1}}{\binom{n+r-1}{r}}.$$

46. (a) There are $(52)_n$ ordered samples of n cards and of these $4 \cdot (48)_{n-1}$ have an ace as the nth card. Thus the required probability is

$$P(A_n) = \frac{4 \cdot (48)_{n-1}}{(52)_n} = \frac{\binom{52-n}{3}}{\binom{52}{4}}, \qquad n = 1, 2, \dots, 49.$$

(b) Let B_n the event that the first ace will appear after the nth card. This is equivalent to the event that the first n cards will contain no ace

$$P(B_n) = \frac{\binom{48}{n}}{\binom{52}{n}}, \qquad n = 0, 1, \dots, 48.$$

Another proof. Let A_n the event in (a). Then $B_n = A_{n+1} \cup A_{n+2} \cup \cdots A_{49}$ and, as the A_j are mutually exclusive,

$$P(B_n) = \sum_{i=n+1}^{49} P(A_i) = \frac{\sum_{i=n+1}^{49} \binom{52-i}{3}}{\binom{52}{4}} = \frac{\binom{52-n}{4}}{\binom{52}{4}} = \frac{\binom{48}{n}}{\binom{52}{n}},$$

where for the sum

$$\sum_{i=n+1}^{49} \binom{52-i}{3}$$

the result of Problem 17(ii) was used.

47. (a) Ignoring leap years, there are 365^v ways of distributing v persons in 365 days. The event A_v that no two persons have the same birthday can be realized in $(365)_v$ ways $(= 0$ for $v > 365)$. Hence

$$P[A_v] = \frac{(365)_v}{365^v} = \prod_{k=1}^{v}\left(1 - \frac{k-1}{365}\right) \approx 1 - \frac{v(v-1)}{730},$$

where the approximation $\approx$ holds for small v. Thus

$$P[A_v'] = 1 - P[A_v] \approx \frac{v(v-1)}{730}.$$

(b) Solving the inequality $P[A_v'] \geq 1/2$ gives $v = 23$ with $P[A_{23}'] = 0.507$. For $v = 60$, it turns out that $P[A_{60}'] = 0.994$, that is, in a class of 60 it is almost certain that at least two have a common birthday.

48. Let A_k be the event that the kth letter is placed in the kth envelope $(k = 1, 2, \ldots, N)$. Then

$$p = P[\text{each letter is placed in a wrong envelope}]$$

$$= 1 - P[\text{at least one letter is placed in the right envelope}]$$

$$= 1 - P\left[\bigcup_{k=1}^{N} A_k\right].$$

By Poincaré's theorem we have

$$P\left[\bigcup_{k=1}^{N} A_k\right] = S_1 - S_2 + \cdots + (-1)^{N-1}S_N, \qquad \text{where}$$

$$S_k = \sum_{1 \leq i_1 < \cdots < i_k \leq N} P[A_{i_1}\ldots A_{i_k}].$$

But

$$P[A_k] = \frac{(N-1)!}{N!}, \qquad S_1 = \sum_{k=1}^{N} P(A_k) = N\frac{(N-1)!}{N!} = 1,$$

$$P[A_iA_j] = \frac{(N-2)!}{N!}, \qquad S_2 = \sum_{i<j} P(A_iA_j) = \binom{N}{2}\frac{(N-2)!}{N!} = \frac{1}{2},$$

$$P(A_iA_jA_k) = \frac{(N-3)!}{N!},$$

so that, in general,

$$P(A_{i_1}A_{i_2},\ldots,A_{i_k}) = \frac{(N-k)!}{N!}, \qquad S_k = \binom{N}{k}\frac{(N-k)!}{N!} = \frac{1}{k!};$$

the last term is

$$S = P(A_1A_2,\ldots,A_N) = \frac{1}{N!}.$$

and so

$$p = 1 - P\left[\bigcup_{k=1}^{N} A_k\right] = 1 - \left[1 - \frac{1}{21} + \frac{1}{31} - \cdots + (-1)^{N-1}\frac{1}{N!}\right] = \sum_{k=2}^{N} (-1)^k \frac{1}{k!}.$$

This probability, even for moderate values of N ($N > 4$), approaches $1 - e^{-1} = 0.632$.

Note: The problem of at least one match (here in terms of envelops and letters) is referred to as the problem of *rencontres*.

49. (a) Let A_k be the event that the number k does not appear ($k = 1;$ $2, \ldots, n$). The probability p_m that exactly m among the n events $A_1, \ldots, A_n$ occur simultaneously is given by

$$p_m = S_m - \binom{m+1}{m}S_{m+1} + \binom{m+2}{m}S_{m+2} - \cdots + (-1)^{n-m}\binom{n}{m}S_n,$$

where S_k is defined as in Problem 48* and is here given by

$$S_k = \binom{n}{k}\frac{(r[n-k])_N}{(rn)_N}.$$

The required probability is given by

$$p_m^* = p_{n-m}.$$

(b) P[each of the numbers $1, 2, \ldots, n$ will appear at least once]

$= 1 - P_n$[at least one of the numbers $1, 2, \ldots, n$ will not appear]

$$= 1 - P\left[\bigcup_{k=1}^{n} A_k\right] = 1 - \sum_{k=1}^{n} (-1)^{k-1}S_k = 1 - \sum_{k=1}^{n} (-1)^{k-1}\binom{n}{k}\frac{([n-k]r)_N}{(nr)_N}$$

$$= \sum_{k=0}^{n} (-1)^k \binom{n}{k}\frac{([n-k]r)_N}{(nr)_N}.$$

50. P[m balls will be needed] $= \binom{n}{1}P$[$n-1$ among the numbers $1, 2,$ $\ldots, n$ will appear at least once after $m-1$ trials, and at the mth trial the nth number will appear]

$$= \binom{n}{1}\left[1 - \sum_{k=1}^{n-1} (-1)^{k-1}\right]S_k \frac{r}{nr - m + 1}$$

$$= \binom{n}{1}\left[\sum_{k=0}^{n-1} (-1)^k \binom{n-1}{k}\left[\frac{([n-k-1]r)_{m-1}}{(nr)_{m-1}}\right]\right] \cdot \frac{r}{nr - m + 1}$$

$$= \sum_{k=1}^{n} (-1)^{k-1} \binom{n-1}{k-1}\frac{([n-k]r)_{m-1}}{(nr-1)_{m-1}}.$$

51. (a) Let A_i be the event that the ith man takes the ith coat and the ith hat. Then

$$p = P[\text{no one takes both his coat and hat}]$$

$$= 1 - P\left[\bigcup_{k=1}^{N} A_k\right]$$

$$= 1 - \sum_{k=1}^{N} (-1)^{k-1} S_k,$$

where

$$S_k = \sum_{1 \leq i_1 < \cdots < i_k \leq N} P[A_{i_1} \ldots A_{i_k}] \quad \text{(Poincaré's theorem)}, \qquad k = 1, \ldots, 2.$$

But

$$P[A_k] = \left[\frac{(N-1)!}{N!}\right]^2, \qquad S_1 = \sum_{k=1}^{N} P(A_k) = N\left[\frac{(N-1)!}{N!}\right]^2 = \frac{(N-1)!}{N! \, 1!},$$

$$P(A_i A_j) = \left[\frac{(N-2)!}{N!}\right]^2, \qquad S_2 = \sum_{1 \leq i < j \leq N} P(A_i A_j) = \binom{N}{2}\left[\frac{(N-2)!}{N!}\right]^2$$

$$= \frac{(N-2)!}{N! \, 2!};$$

and in general

$$P[A_{i_1} A_{i_2} \ldots A_{i_k}] = \left[\frac{(N-k)!}{N!}\right]^2, \qquad S_k = \binom{N}{k}\left[\frac{(N-k)!}{N!}\right]^2 = \frac{(N-k)!}{N! \, k!};$$

the last term is

$$S_N = P(A_1 A_2 \ldots A_N) = \left(\frac{1}{N!}\right)^2.$$

Therefore

$$p = 1 - \sum_{k=1}^{N} (-1)^{k-1} \frac{(N-k)!}{N! \, k!} = \sum_{k=0}^{N} (-1)^k \frac{(N-k)!}{N! \, k!}.$$

(b) $P[\text{each man takes the wrong coat and the wrong hat}]$

$$= P[\text{each takes a wrong coat}] \cdot P[\text{each takes a wrong hat}]$$

$$= \left[\sum_{k=2}^{N} (-1)^k \frac{1}{k!}\right]\left[\sum_{k=2}^{N} (-1)^k \frac{1}{k!}\right] = \left[\sum_{k=2}^{N} (-1)^k \frac{1}{k!}\right]^2 \quad \text{(see Exercise 48).}$$

52. (a) Let p_k be the probability that the housewife gets k free packets. Then

$$p_0 = P[A_1 \cup A_2 \cup A_3 A_4] = S_1 - S_2 + S_3 - S_4 \tag{1}$$

with S_i ($i = 1, 2, 3, 4$), defined as in Problem 48*, and let A_i be the event that the ith letter of the word TIDE does not appear, $(i = 1, 2, 3, 4)$. We have

$$P(A_i) = \left(\frac{3}{4}\right)^8, \qquad P(A_iA_j) = \left(\frac{1}{2}\right)^8, \qquad P(A_iA_jA_k)$$

$$= \left(\frac{1}{4}\right)^8, \qquad \text{and} \qquad P(A_1A_2A_3A_4) = 0.$$

Substituting into (1) we get

$$p_0 = 4\left(\frac{3}{4}\right)^8 - \binom{4}{2}\left(\frac{1}{2}\right)^8 + \binom{4}{3}\left(\frac{1}{4}\right)^8.$$

On the other hand,

$$p_2 = \frac{8!}{(2!)^4}\left(\frac{1}{4}\right)^8 \qquad \text{(multinomial distribution).} \tag{2}$$

53. Let A be the event that both the selected balls are white. We have:

$$P(A) = \frac{N_1}{N_1 + N_2} \cdot \frac{N_1 - 1}{N_1 + N_2 - 1} = \frac{1}{2}, \tag{1}$$

and because of

$$\frac{N_1}{N_1 + N_2} > \frac{N_1 - 1}{N_1 + N_2 - 1} \qquad \text{for} \quad N_2 > 0,$$

equation (1) gives the inequalities

$$\left(\frac{N_1}{N_1 + N_2}\right)^2 > \frac{1}{2} > \left(\frac{N_1 - 1}{N_1 + N_2 - 1}\right)^2,$$

the first of which gives

$$N_1 > \frac{1}{\sqrt{2}}(N_1 + N_2), \qquad \text{that is,} \quad N_1 > (\sqrt{2} + 1)N_2,$$

and the second

$$(1 + \sqrt{2})N_2 > N_1 - 1.$$

Hence, it follows that

$$(1 + \sqrt{2})N_2 < N_1 < (1 + \sqrt{2})N_2 + 1. \tag{2}$$

For $N_2 = 1$ we get $2.41 < N_1 < 3.14$, that is, $N_1 = 3$. Then we observe that

$$P(A) = \frac{3}{4} \cdot \frac{2}{3} = \frac{1}{2}$$

and the required minimum value of N_1 is 3.

(b) If N_2 is even, then, from (1) and (2), we have

N_2	N_1 between	possible N_1	$P(A)$
2	4.8 and 5.8	5	10/21
4	9.7 and 10.7	10	45/91
6	14.5 and 15.5	15	1/2

Thus the minimum value of N_1 is 15.

(c) The minimum value of N corresponds to the minimum value of N_1 since N_1 is an increasing function of N_2. Thus in (a) $N = 3 + 1 = 4$, and in (b) $N = 15 + 6 = 21$.

54. Since A and B are independent we have

$$P(AB) = P(A)P(B),$$

and hence

$$P(AB') = P(A) - P(AB) = P(A) - P(A)P(B) = P(A)[1 - P(B)]$$
$$= P(A)P(B'), \tag{1}$$

which shows that A and B' are independent. Similarly, it is shown that

$$P(A'B) = P(A')P(B), \tag{2}$$

$$P(A'B') = P(A \cup B)' = 1 - P(A \cup B) = 1 - P(A) - P(B) + P(AB)$$
$$= 1 - P(A) - P(B) + P(A)P(B) = [1 - P(A)][1 - P(B)]$$
$$= P(A')P(B').$$

55. Since the events A, B, and C are completely independent, the following relations hold:

$$P(AB) = P(A) \cdot P(B), \tag{1}$$

$$P(BC) = P(B)P(C), \tag{2}$$

$$P(CA) = P(C)P(A), \tag{3}$$

$$P(ABC) = P(A) \cdot P(B)P(C). \tag{4}$$

From (2) and (4) we obtain

$$P(AA^*) = P(ABC) = P(A) \cdot P(BC) = P(A) \cdot P(A^*),$$

which shows that the events A and $A^* = BC$ are independent. The proof of the remainder is similar.

56. $x = P(A'B'C) = P(A \cup B)'C = P(C) - P(A \cup B)C = P(C)$

$$- P(AC \cup BC)$$

$$= P(C) - P(AC) - P(BC) + P(ABC)$$

$$= P(C) - P(A)P(C) - P(B)P(C) + P(A)P(B)P(C).$$

Letting $P(A) = a$ we obtain

$$x = (1 - a)(1 - P(B))P(C), \tag{1}$$

$$b = 1 - P(A \cup B \cup C) = P(A'B'C') = P(A')P(B')P(C')$$

$$= (1 - P(A))(1 - P(B))(1 - P(C)).$$

$$b = (1 - a)(1 - P(B))(1 - P(C)), \tag{2}$$

$$c = 1 - P[ABC] = 1 - P(A)P(B)P(C) = 1 - aP(B)P(C). \tag{3}$$

From (1) and (2) we get

$$\frac{x}{b} = \frac{P(C)}{1 - P(C)} \quad \Rightarrow \quad P[C] = \frac{x}{x + b}. \tag{4}$$

Substituting (4) into (3) we obtain

$$P(B) = \frac{(1 - c)(x + b)}{ax}. \tag{5}$$

Combining (1), (4), and (5) we get

$$x = (1 - a)\left(1 - \frac{(1 - c)(x + b)}{ax}\right) \cdot \frac{x}{x + b} = \frac{1 - a}{x + b} \cdot \frac{ax - (1 - c)(x + b)}{a},$$

$$ax(x + b) = (1 - a)ax - (1 - a)(1 - c)(x + b),$$

$$ax^2 + abx = (1 - a)ax - (1 - a)(1 - c)x - (1 - a)(1 - c)b,$$

$$ax^2 + ab - (1 - a)(a + c - 1)x + (1 - a)(1 - c)b = 0. \tag{6}$$

As x represents probability, both roots of (6) must be positive and therefore their sum as well, i.e.,

$$(1 - a)(a + c - 1) - ab > 0, \qquad a - 1 + c > \frac{ab}{1 - a}, \qquad c > \frac{(1 - a)^2 + ab}{1 - a}.$$

57. Associate the events A_i with independent Bernoulli trials; hence use the binomial distribution.

58. Let A_i be the event that an ace appears at the ith throw and let B_i be the event that a six appears at the ith throw. Then the required probability equals

$$P[B_1' B_2' | A_1' A_2'] = \frac{P[A_1' B_1' A_2' B_2']}{P[A_1' A_2']} = \frac{P(A_1' B_1')P(A_2' B_2')}{P(A_1')P(A_2')}$$

$$= \frac{\left(\frac{4}{6}\right)^2}{\left(\frac{5}{6}\right)^2} = \left(\frac{4}{5}\right)^2 = \frac{16}{25}.$$

59. (a) We have

$$P[X_2 > 0] = 1 - P[X_2 = 0].$$

By the theorem of total probability

$$P[X_2 = 0] = P[X_1 = 0]P[X_2 = 0|X_1 = 0] + P[X_1 = 1]P[X_2 = 0|X_1 = 1]$$

$$+ P[X_1 = 2]P[X_2 = 0|X_1 = 2] = \frac{1}{4} \cdot 1 + \frac{1}{2} \cdot \frac{1}{4} + \frac{1}{4} \cdot \frac{1}{16} = \frac{25}{64},$$

and hence

$$P[X_2 > 0] = 1 - \frac{25}{64} = \frac{39}{64}.$$

(b) Using Bayes's formula, the required probability equals

$$P[X_1 = 2|X_2 = 1]$$

$$= \frac{P[X_1 = 2]P[X_2 = 1|X_1 = 2]}{P[X_1 = 1]P[X_2 = 1|X_1 = 1] + P[X_1 = 2]P[X_2 = 1|X_1 = 2]}$$

$$= \frac{\dfrac{1}{4} \cdot \dfrac{1}{4}}{\dfrac{1}{2} \cdot \dfrac{1}{2} + \dfrac{1}{4} \cdot \dfrac{1}{4}} = \frac{1}{5}.$$

60. Let A_m denote the event $X_{max} \leq m$ $(m = 1, 2, \ldots, n)$. Then

$$P[A_n] = P[X_1 \leq m, X_2 \leq m, \ldots, X_r \leq m]$$

$$= P[X_1 \leq m] \cdots P[X_r \leq m] = \frac{m^r}{n^r}.$$

Clearly, $A_{m-1} A_m$ and the event $[X_{max} = m] = A_m - A_{m-1}$. Consequently,

(a) $P[X_{max} = m] = P[A_m] - P[A_{m-1} A_m] = P[A_m] - P[A_{m-1}]$

$$= \frac{m^r - (m-1)^r}{n^r},$$

(b) $P[X_{max} = m] = \dfrac{\dbinom{m-1}{r-1}}{\dbinom{n}{r}}, \qquad m = r, r+1, \ldots, n.$

61. The number X of successes (wins) of a team in n games obeys a binomial distribution with probability $p = 1/2$ for success, that is,

$$P[X = k] = \binom{n}{k}\left(\frac{1}{2}\right)^n, \qquad k = 0, 1, 2, \ldots, n.$$

(a) $P[\text{the series will end in at most 6 games}]$

$$= P[\text{in exactly 4 games}] + P[\text{in exactly 5 games}]$$

$$+ P[\text{in exactly 6 games}].$$

Now we have

$$P[\text{in exactly 4 games}] = P[A \text{ wins the 4 games}] + P[B \text{ wins the 4 games}]$$

$$= 2 \cdot \binom{3}{3}\left(\frac{1}{2}\right)^4 = \frac{1}{8}.$$

$P[\text{in exactly 5 games}] = 2 \cdot P[A \text{ wins 3 of the first 4 games and also wins}$

$$\text{the 5th (last) game}] = 2 \cdot \binom{4}{3}\left(\frac{1}{2}\right)^5 = \frac{1}{4}.$$

$$P[\text{in exactly 6 games}] = 2 \cdot \binom{5}{3}\left(\frac{1}{2}\right)^6 = \frac{5}{16}.$$

Hence,

$$P[\text{the series will end in at most 6 games}] = \frac{1}{8} + \frac{1}{4} + \frac{5}{16} = \frac{11}{16}.$$

(b) The required probability p equals

$$p = P[B \text{ will win the games 3 to 6}]$$

$$+ P[A \text{ will win 2 of the games 3 to 5 and the sixth game}]$$

$$= (\tfrac{1}{2})^4 + 3(\tfrac{1}{2})^4 = \tfrac{1}{4}.$$

62. Let A_i be the event when the patient has illness A_i ($i = 1, 2, 3$), and let B be the event when the result of the test is positive twice. By hypothesis, the *a priori* probabilities are

$$P(A_1) = \frac{1}{2}, \qquad P(A_2) = \frac{1}{4}, \qquad P(A_3) = \frac{1}{4}.$$

After the three independent repetitions of the test the probabilities that the result will be positive are

$$P[B|A_1] = \binom{3}{2}\left(\frac{1}{4}\right)^2 \frac{3}{4} = \frac{9}{64},$$

$$P[B|A_2] = \binom{3}{2}\left(\frac{1}{2}\right)^3 = \frac{3}{8},$$

$$P[B|A_3] = \binom{3}{2}\left(\frac{3}{4}\right)^2 \frac{1}{4} = \frac{27}{64}.$$

From Bayes's formula we get the required probabilities

$$P[A_1|B] = \frac{\dfrac{2}{4} \cdot \dfrac{9}{64}}{\dfrac{2}{4} \cdot \dfrac{9}{64} + \dfrac{1}{4} \cdot \dfrac{24}{64} + \dfrac{1}{4} \cdot \dfrac{27}{64}} = \frac{18}{18 + 24 + 27} = \frac{18}{69},$$

$$P[A_2|B] = \frac{24}{68}, \qquad P[A_3|B] = \frac{27}{69}.$$

63. Let G be the event that a person is Greek, let T be the event that a person is Turkish, and let E be the event that a person speaks English. Then from Bayes's formula we have

$$P[G|E] = \frac{P[G]P[E|G]}{P[G]P[E|G] + P[T]P[E|T]} = \frac{0.75 \times 0.20}{0.75 \times 0.20 + 0.25 \times 0.10}$$

$$= 0.857,$$

that is, of the English speaking population of Nicosia, 85.7% are Greeks.

64. Let A_i be the event that A forgets his umbrella in the ith shop, let B_i be the event that B forgets his umbrella in the ith shop, and let B_0 be the event that B has left his umbrella at home. Then:

(a) $P(\text{they have both umbrellas}) = P(A_1' A_2' A_3')[P(B_0) + P(B_0' B_1' B_2' B_3')]$

$$= \left(\frac{3}{4}\right)^3 \left[\frac{1}{2} + \frac{1}{2}\left(\frac{3}{4}\right)^3\right]$$

$$= \frac{27}{64} \cdot \frac{64 + 27}{2 \cdot 64} = \frac{2457}{8192}.$$

(b) $P(\text{they have only one umbrella})$

$$= P(A_1' A_2' A_3')[P(B_0' B_1) + P(B_0' B_1' B_2) + P(B_0' B_1' B_2' B_3')]$$

$$+ [P(A_1) + P(A_1' A_2) + P(A_1' A_3')][P(B_0) + P(B_0' B_1' B_2' B_3')]$$

$$= \left(\frac{3}{4}\right)^3 \left[\frac{1}{2}\frac{1}{4}\frac{1}{2}\frac{3}{4}\frac{1}{4} + \frac{1}{2}\left(\frac{3}{4}\right)^2\frac{1}{4}\right]$$

$$+ \left[\frac{3}{4} + \frac{1}{4}\frac{3}{4} + \left(\frac{3}{4}\right)^2\frac{1}{4}\right] \cdot \left[\frac{1}{2} + \frac{1}{2}\left(\frac{3}{4}\right)^3\right]$$

$$= \frac{999}{8192} + \frac{6279}{8192} = \frac{7278}{8192}$$

(c) $P(B \text{ lost his umbrella}|\text{they have only one umbrella})$

$$= \frac{P(\text{they have one umbrella and } B \text{ lost his umbrella})}{P(\text{they have only one umbrella})}$$

$$= \frac{\dfrac{999}{8192}}{\dfrac{7278}{8192}} = \frac{999}{7218}.$$

65. The condition for A and B to be independent is

$$P(AB) = P(A)P(B). \qquad (*)$$

We have

$$P(A) = 1 - P(A') = 1 - [P(\text{all boys}) + P(\text{all girls})]$$

$$= 1 - \left(\frac{1}{2^n} + \frac{1}{2^n}\right) = \frac{2^{n-1} - 1}{2^{n-1}},$$

$$P(B) = P(\text{all boys}) + P(\text{one girl}) = \left(\frac{1}{2}\right)^n + n\left(\frac{1}{2}\right)\left(\frac{1}{2}\right)^{n-1} = \frac{n+1}{2^n},$$

$$P(AB) = P(\text{one girl}) = \frac{n}{2^n},$$

and by condition (∗) we must have

$$\frac{n}{2^n} = \frac{n+1}{2^n}\frac{2^{n-1} - 1}{2^n - 1} \Rightarrow 2^{n-1} = n + 1 \Rightarrow n = 3.$$

66. Let A be the event that a really able candidate passes the test and let B be the event that any candidate passes the test. Then we have

$$P[B|A] = 0.8, \qquad P[B|A'] = 0.25, \qquad P(A) = 0.4, \qquad P(A') = 0.6,$$

and by Bayes's formula

$$P[A|B] = \frac{P(A)P(B|A)}{P(A)P(B|A) + P(A')P(B|A')} = \frac{0.32}{0.32 + 0.15} = \frac{32}{47},$$

that is, about 68%.

67. $$P[A] = \sum_{k=0}^{\infty} P[A \text{ wins at the } k + 1 \text{ throw of the dice}]$$

$$= \sum_{k=0}^{\infty} q_1^k q_2^k p_1 = \frac{p_1}{1 - q_1 q_2},$$

where

$$p_1 = P[A \text{ wins in a throw of the dice}] = \frac{5}{36}, \qquad q_1 = 1 - p_1 = \frac{31}{36},$$

$$p_2 = P[B \text{ wins in a throw of the dice}] = \frac{6}{36}, \qquad q_2 = 1 - p_2 = \frac{30}{36};$$

consequently,

$$P(A) = \frac{\dfrac{5}{36}}{1 - \dfrac{31}{36}\cdot\dfrac{30}{36}} = \frac{30}{61}.$$

68. Let A_i be the event that player P_i wins ($i = 1, 2, 3$), let B_j be the event that a non-ace appears at the jth throw, and let $P(A_i) = a_i$ $(i = 1, 2, 3)$. Then

$$A_2 \subset B_1 \qquad \text{and} \qquad A_3 \subset B_1 B_2;$$

so $A_2 = A_2 B_1$, $A_3 = A_3 B_1 B_2$, and

$$a_2 = P(A_2) = P(A_2 B_1) = P(B_1)P(A_2|B_1) = \frac{5}{6} a_1,$$

because if P_1 does not win on the first throw then P_2 plays "first" and hence $P(A_2|B_1) = P(A_1) = a_1$. Similarly,

$$a_3 = P(A_3) = P(A_3 B_1 B_2) = P(B_1 B_2)P(A_3|B_1 B_2) = \frac{25}{36} a_1,$$

and, since $a_1 + a_2 + a_3 = 1$, we get

$$a_1 = \frac{36}{91}, \qquad a_2 = \frac{30}{91}, \qquad a_3 = \frac{25}{91}.$$

69. By an argument similar to Exercise 68 we get

$$p_k = q^{k-1} p_1, \qquad k = 1, 2, \dots, N, \quad q = 1 - p.$$

and by the condition $\sum_{k=1}^{N} p_k = 1$

$$p_k = \frac{pq^{k-1}}{1 - q^N}, \qquad k = 1, 2, \dots, N.$$

70. Let W be the event that finally a white ball is selected from A, and W_1 be the event that a white ball is selected from A, and let W_2 be the event that a white ball is selected from B. Then by the theorem of total probability we get

$$P[W] = P[W|W_1 W_2] \cdot P[W_1 W_2] + P[W|W_1 W_2']P[W_1 W_2']$$
$$+ P[W|W_1' W_2]P[W_1' W_2] + P[W|W_1' W_2']P[W_1' W_2'],$$

where setting $w_1 + b_i = N_i$ $(i = 1, 2)$, we have

$$P[W|W_1 W_2] = \frac{w_1}{N_1}, \qquad\qquad P[W_1 W_2] = \frac{w_1}{N_1} \cdot \frac{w_2 + 1}{N_2 + 1},$$

$$P[W|W_1 W_2'] = \frac{w_1 - 1}{N_1}, \qquad\quad P[W_1 W_2'] = \frac{w_1}{N_1} \cdot \frac{b_2}{N_2 + 1},$$

$$P[W|W_1 W_2] = \frac{w_1 - 1}{N_1}, \qquad\quad P[W_1' W_2] = \frac{b_1}{N_1} \cdot \frac{w_2}{N_2 + 1},$$

$$P[W|W_1' W_2'] = \frac{w_1}{N_1}, \qquad\qquad P[W_1' W_2'] = \frac{b_1}{N_1} \cdot \frac{b_2 + 1}{N_2 + 1}.$$

71. (a) Let A be the event that a person has the disease, and let B_i be the event that the ith test is positive ($i = 1, 2$). Then

$$P[\overset{\iota}{A}] = 0.1, \qquad P[B_i|A] = 0.9, \qquad P[B_i|A'] = 1 - P[B_i'|A'] = 0.1.$$

Letting $B = B_1 B_2$ we have

$$P[B|A] = P[B_1|A]P[B_2|A] = 0.81, \qquad P[B|A'] = 0.01,$$

$$P[A|B] = \frac{P[A] \cdot P[B|A]}{P[A] \cdot P[B|A] + P[A'] \cdot P[B|A']} = \frac{81}{81 + 9} = 0.9.$$

(b) If C denotes the event that only one test is positive then

$$P[C|A] = \binom{2}{1} P[B_i|A] \cdot P[B_i'|A] = 0.18,$$

$$P[C|A'] = \binom{2}{1} P[B_1|A'] \cdot P[B_i'|A'] = 0.18.$$

Thus we find

$$P[A|C] = \frac{P[A]P[C|A]}{P[A] \cdot P[C|A] + P[A']P[C|A']} = 0.1.$$

72. (a) Let A be the event that coin A is selected, let B be the event that coin B is selected, and let H_k be the event that heads appears k times ($k = 0, 1, 2$). Then by the theorem of total probability

$$P[H_k] = P[A]P[H_k|A] + P[B]P[H_k|B].$$

We have

$$P[A] = P[B] = \frac{1}{2}, \qquad P[H_k|A] = \binom{2}{k}\left(\frac{3}{4}\right)^k\left(\frac{1}{4}\right)^{2-k},$$

$$P[H_k|B] = \binom{2}{k}\left(\frac{1}{4}\right)^k\left(\frac{3}{4}\right)^{2-k},$$

and therefore

(i) $$P[H_2] = \frac{1}{2}\left[\left(\frac{3}{4}\right)^2 + \left(\frac{1}{4}\right)^2\right] = \frac{5}{16}.$$

(ii) $$P[H_1] = \frac{1}{2}\left[2 \times \frac{3}{4} \times \frac{1}{4} + 2 \times \frac{1}{4} \times \frac{3}{4}\right] = \frac{3}{8}.$$

(b) P[heads at least once in strategy (a)] $= P[H_1] + P[H_2]$

$$= \frac{11}{16},$$

P[heads at least once in strategy (b)] $= \binom{2}{1}\left(\frac{1}{2}\right)^2 + \binom{2}{2}\left(\frac{1}{2}\right)^2$

$$= \frac{3}{4}.$$

Hence he must follow the second strategy if he wants to maximize the probability of at least one head.

73. Let A be the event that a white ball is selected from A, let B_1 be the event that two white balls are selected from B, let B_2 be the event that one white ball and one black ball are selected from B, and let B_3 be the event that two black balls are selected from B. Then by the theorem of total probability we get:

(a) $\qquad P[A] = P[B_1]P[A|B_1] + P[B_2]P[A|B_2] + P[B_3]P[A|B_3],$

$$P[B_1] = \frac{\binom{4}{2}}{\binom{12}{2}}, \qquad P[B_2] = \frac{\binom{8}{1}\binom{4}{1}}{\binom{12}{2}}, \qquad P[B_3] = \frac{\binom{8}{3}}{\binom{12}{2}},$$

$$P[A|B_1] = \frac{8}{13}, \qquad P[A|B_2] = \frac{7}{13}, \qquad P[A|B_3] = \frac{6}{13}.$$

Hence

$$P[A] = \frac{136}{429}.$$

(b) The required probability p say, is given by

$$p = P[B_1|A] + P[B_2|A].$$

Applying Bayes's formula we get

$$P[B_1|A] = \frac{P[B_1]P[A|B_1]}{P[A]} = \frac{6}{34}, \qquad P[B_2|A] = \frac{P[B_2]P[A|B_2]}{P[A]} = \frac{7}{34}.$$

Hence

$$p = \frac{13}{34}.$$

74. Let the event A be that the test is positive and the event B that a woman has the disease. Then

$$P[B|A] = \frac{P(B)P(A|B)}{P(B)P(A|B) + P(B')P(A|B')} = \frac{\dfrac{1}{2000} \cdot \dfrac{19}{20}}{\dfrac{1}{2000} \cdot \dfrac{19}{20} + \dfrac{1999}{2000} \cdot \dfrac{1}{20}} = \frac{19}{2018},$$

that is, less than 1% of the women with positive test have the illness. The lady's fear is rather unjustified.

75. Let A, B, C be the events that the clerk follows routes A, B, C, respectively, let L be the event that the clerk arrives late, and let S be the event that the day is sunny. Then

(a) $P[C|SL] = \dfrac{P[CSL]}{P[SL]} = \dfrac{P[S]P[CL|S]}{P[S]P[L|S]}$

$= \dfrac{P[C|S]P[L|CS]}{P[A|S]P[L|AS] + P[B|S]P[L|BS] + P[C|S]P[L|CS]}.$

Since

$$P[A|S] = P[B|S] = P[C|S] = \frac{1}{3}, \qquad P[L|AS] = 0.05,$$

$$P[L|BS] = 0.10, \qquad P[L|CS] = 0.15,$$

we get $P[C|SL] = 0.5$.

(b) $P[S'|L] = \dfrac{P[S']P[L|S]}{P[S]P[L|S] + P[S']P[L|S']},$

$P[L|S] = P[A|S]P[L|AS] + P[B|S]P[L|BS]$

$$+ P[C|S]P[L|CS] = \frac{1}{3}\frac{30}{100},$$

$P[L|S'] = P[A|S']P[L|AS'] + P[L|BS'] + P[C|S']P[L|CS']$

$$= \frac{1}{3}\left(\frac{6}{100} + \frac{15}{100} + \frac{20}{100}\right) = \frac{1}{3}\cdot\frac{41}{100},$$

$$P(S) = \frac{3}{4}, \qquad P(S') = \frac{1}{4}.$$

Consequently,

$$P[S'|L] = \frac{41}{131}.$$

76. Let A be the event that the painting is original, and let B be the event that the expert says it is an original. Then according to Bayes's formula

(a) $P[A|B] = \dfrac{P[A]P[B|A]}{P[A]P[B|A] + P[A']P[B|A']} = \dfrac{\dfrac{5}{6}\cdot\dfrac{9}{10}}{\dfrac{5}{6}\cdot\dfrac{9}{10} + \dfrac{1}{6}\cdot\dfrac{1}{10}} = \dfrac{45}{46}.$

(b) Let A be the event that the second choice is original, and let H be the event that the expert decides correctly that it is an original. Then by the total probability theorem we have

$$P[A^*] = P[A^*|H]P[H] + P[A^*|H']P[H'] = \frac{9}{10}\cdot\frac{10}{11} + \frac{1}{10}\cdot\frac{9}{11} = \frac{99}{110}.$$

77. Let A be the event that the 5 cards are clubs, and let B be the event that the 5 cards are black. Then

$$P[A|B] = \frac{P[AB]}{P[B]} = \frac{\binom{13}{5}/\binom{52}{5}}{\binom{26}{5}/\binom{52}{5}} = \frac{\binom{13}{5}}{\binom{26}{5}}.$$

78. Let A_2 denote 2 boys, let B_2 denote 2 girls, and let C_n denote n children in the family. Then

(a)
$$P[C_2|A_2] = \frac{P[C_2]P[A_2|C_2]}{\sum\limits_{n=2}^{\infty} P[C_n]P[A_2|C_n]}.$$

(b)
$$P[B_2|A_2] = \frac{P[A_2B_2]}{P[A_2]} = \frac{\sum\limits_{n=4}^{\infty} P[C_n]P[A_2B_2|C_n]}{\sum\limits_{n=2}^{\infty} P[C_n]P[A_2|C_n]},$$

and since

$$p_n = P[C_n] = (1 - 2a)\left(\frac{1}{2}\right)^{n-1}, \qquad n = 2, 3, \ldots,$$

and

$$P[A_2|C_n] = \binom{n}{2}\left(\frac{1}{2}\right)^n,$$

we get

$$P[A_2] = \sum_{n=2}^{\infty} P[C_n] \cdot P[A_2|C_n] = (1 - 2a)\sum_{n=2}^{\infty}\binom{n}{2}\left(\frac{1}{2}\right)^{2n-1}$$

$$= \frac{(1 - 2a)}{4^2}\sum_{n=2}^{\infty} n(n-1)\left(\frac{1}{4}\right)^{n-2} = \frac{(1 - 2a)}{4^2}\frac{2}{\left(1 - \frac{1}{4}\right)^3} = \frac{8(1 - 2a)}{27},$$

$$P[A_2B_2] = \sum_{n=4}^{\infty} P[C_n] \cdot P[A_2B_2|C_n] = P[C_4] \cdot P[A_2B_2|C_4]$$

$$= (1 - 2a)\left(\frac{1}{2}\right)^3\binom{4}{2}\left(\frac{1}{2}\right)^4$$

$$= \frac{6(1 - 2a)}{128} = \frac{3(1 - 2a)}{64}.$$

Finally,

$$P[C_2|A_2] = \frac{\dfrac{(1 - 2a)}{8}}{\dfrac{8(1 - 2a)}{27}} = \frac{27}{64},$$

$$P[B_2|A_2] = \dfrac{\dfrac{3(1-2a)}{64}}{\dfrac{8(1-2a)}{27}} = \dfrac{81}{512}.$$

79. (a)

$$P[A_3|B_3] = \dfrac{P[A_3] \cdot P[B_3|A_3]}{P[A_1]P[B_3|A_1] + P[A_2]P[B_3|A_2] + P[A_3]P[B_3|A_3]}$$

$$= \dfrac{6}{11}.$$

(b)

$$P[A_2|B_3] = \dfrac{P[A_2]P[B_3|A_2]}{\sum\limits_{i=1}^{3} P[A_i]P[B_3|A_i]} = \dfrac{0.2 \times 0.1}{0.11} = \dfrac{2}{11}.$$

(c)

$$P[A_2|B_1 \cup B_2] = \dfrac{P[A_2] \cdot P[B_1 \cup B_2|A_2]}{\sum\limits_{i=1}^{3} P[A_i] \cdot P[B_1 \cup B_2|A_i]},$$

whereby the addition theorem for conditional probabilities

$$P[B_1 \cup B_2|A_i] = P[B_1|A_i] + P[B_2|A_i] - P[B_1 B_2|A_i],$$

and the conditional probabilities are given in the table.

80.

(Ia) $p_k = \dbinom{5}{k}\left(\dfrac{7}{10}\right)^k\left(\dfrac{3}{10}\right)^{5-k}$, $k = 0, 1, \ldots, 5$ (binomial).

(Ib) $p_k = \dfrac{\dbinom{7}{k}\dbinom{3}{5-k}}{\dbinom{10}{5}}$, $k = 2, 3, 4, 5$ (hypergeometric).

(IIa) $P[X_{\min} > k] = \dfrac{(10-k)^5}{10^5}$,

$$P[X_{\min} = k] = P[X_{\min} > k - 1] - P[X_{\min} > k]$$

$$= \dfrac{(10-k+1)^5 - (10-k)^5}{10^5}, \qquad k = 1, 2, \ldots, 10.$$

(IIb) $P[X_{\min} > k] = \dfrac{(10-k)_5}{(10)_5}$

(IIIa) $P[X_{max} \leq k] = \left(\dfrac{k}{10}\right)^5,$

$$P[X_{max} = k] = P[X_{max} \leq k] - P[X_{max} \leq k - 1]$$

$$= \frac{k^5 - (k - 1)^5}{10^5}, \qquad k = 1, 2, \ldots, 10.$$

(IIIb) $P[X_{max} \leq k] = \dfrac{(k)_5}{(10)_5} \qquad k = 5, 6, \ldots, 10.$

(IVa) $p_k = \left(\dfrac{7}{10}\right)\left(\dfrac{3}{10}\right)^{k-1}, \qquad k = 1, 2, \ldots.$

(IVb) $p_k = \dfrac{7(3)_{k-1}}{(10)_k}, \qquad k = 1, 2, 3, 4.$

81. (I) and (IV) as in the preceding problem.

(IIa) $P[X_{min} \geq k] = \begin{cases} \left[\dfrac{10 - 2(k - 1)}{10}\right]^5, & k = 1, 2, 3, \\[3mm] \left[\dfrac{7 - (k - 1)}{10}\right]^5, & k = 4, 5, 6, 7. \end{cases}$

(IIb) $P[X_{min} \geq k] = \begin{cases} \dfrac{10 - 2(k - 1)_5}{(10)_5}, & k = 1, 2, 3, \\[3mm] \dfrac{(7 - (k - 1))_5}{(10)_5}, & k = 4, 5, 6, 7. \end{cases}$

(IIIa) $P[X_{max} \leq k] = \begin{cases} (2k/10)^5, & k = 1, 2, 3, \\ (k + 3)/(10)^5, & k = 4, 5, 6, \end{cases}$

(IIIb) $P[X_{max} \leq k] = \begin{cases} (2k)_5/(10)_5, & k = 1, 2, 3, \\ (k + 3)_5/(10)_5, & k = 4, 5, 6, 7. \end{cases}$

(IIIc) $P[X_{max} \leq k] = (2k/10)^5, \qquad k = 1, 2, 3,$

$$P[X_{max} \leq k] = (k + 3/10), \qquad k = 4, 5, 6, 7.$$

82. Let X be the number of defective items in a sample of 10. Then X obeys a binomial law

$$P(X = k) = \binom{10}{k}(1/10)^k(9/10)^{10-k}, \qquad k = 0, 1, \ldots, 10,$$

and the required probability is

$$P(X = 0) = (9/10)^{10}.$$

83. Let X be the number of persons that incur the accident in a year. Then X obeys a binomial law with $n = 5{,}000$ and $p = 1/1{,}000$. Since $\lambda = np = 5$, the

Poisson approximation to the binomial (for large n and small p) applies. Therefore, the required probability is given by

$$P(X \le 2) = e^{-5}\left(\frac{5^0}{0!} + \frac{5}{1!} + \frac{5^2}{2!}\right) = \frac{31}{2e^{-5}}.$$

84. Let X be the number of persons making reservations on a flight and not showing up for the flight. Then the random variable X has the binomial distribution with $n = 100$ and $p = 0.05$. Hence

$$P(X \le 95) = 1 - P(X > 95) = 1 - \sum_{k=96}^{100} \binom{100}{k}(0.05)^k(0.95)^{100-k}.$$

85. (i) Let X be the number of accidents per week. Then X obeys a Poisson probability law with parameter $\lambda = 2$ and hence

$$P(X \le 2) = e^{-2}(1 + 2/1! + 2^2/2!) = 5e^{-2}.$$

(ii) If Y denotes the number of accidents in 2 weeks, then Y has A Poisson distribution with $\lambda = 4$. Hence

$$P(Y \le 2) = e^{-4}(1 + 4/1! + 4^2/2!) = 13e^{-4}.$$

(iii) The required probability equals $(P[X \le 2])^2 = 25e^{-4}$.

86. The number X of suicides per month obeys a binomial probability law with parameters $n = 5 \times 10^5$ and $p = 4 \times 10^{-6}$. Since $np = 2$ (< 10) the Poisson probability law with $\lambda = 2$ provides a satisfactory approximation to the above binomial law. Hence,

$$p_0 \equiv P(X \le 4) = \sum_{k=0}^{4} e^{-2}\frac{2^k}{k!} = 7e^{-2}.$$

Let Y be the number of months with more than four suicides. Then

$$P(Y = k) = \binom{12}{k}(1 - p_0)^k(p_0)^{12-k},$$

$$P(Y \ge 2) = 1 - \{P(X = 0) + P(X = 1)\}.$$

87. Because of the independence of the tosses the required probability is equal to the probability that 20 tosses result in n heads, that is,

$$\binom{20}{n}\frac{1}{2^{20}}.$$

The required conditional probability equals

$$\frac{\binom{30}{10+n}\left(\frac{1}{2}\right)^{30}}{\sum_{k=0}^{20}\binom{30}{10+k}\left(\frac{1}{2}\right)^{30}} = \frac{\binom{30}{10+n}}{\sum_{k=0}^{20}\binom{30}{10+k}}.$$

88. Suppose that a family has n children and let X be the number of boys. Then X obeys a binomial law with parameters n and $p = 1/2$. The probability of at least one boy and at least one girl equals

$$P(1 \le X \le n - 1) = 1 - P(X = 0) - P(X = n) = 1 - 2\frac{1}{2^n} = 1 - \frac{1}{2^{n-1}},$$

and since this probability must be (at least) 0.95 we get $2^{n-1} \ge 20$ and hence $n = 6$.

89. $$p(k; \lambda) = e^{-\lambda}\frac{\lambda^k}{k!} = \frac{\lambda}{k}e^{-\lambda}\frac{\lambda^{k-1}}{(k-1)!} = \frac{\lambda}{k}p(k - 1; \lambda).$$

For given λ, we conclude from the above relation that $p(k; \lambda)$ is increasing in k for $k < \lambda$ and decreasing for $k > \lambda$. If λ happens to be an integer then $p(\lambda; \lambda) = p(\lambda - 1; \lambda)$. Thus $p(k; \lambda)$ attains its maximum when k is the largest integer not exceeding λ.

90. (a) $$\int_{-\infty}^{\infty} f(x)\,dx = \int_0^2 (1 - |1 - x|)\,dx$$

$$= \int_0^1 x\,dx + \int_1^2 (2 - x)\,dx = \frac{1}{2} + \frac{1}{2} = 1.$$

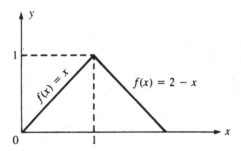

Triangular distribution

(b) $$\frac{1}{\pi}\int_{-\infty}^{\infty} \frac{\beta\,dx}{\beta^2 + (x - a)^2} = \frac{1}{\pi}\left[\arctan\left(\frac{x - a}{\beta}\right)\right]_{-\infty}^{\infty} = 1.$$

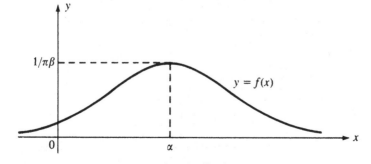

Cauchy distribution

(c) $$\int_{-\infty}^{\infty} f(x)\, dx = \frac{1}{2\sigma} \int_{-\infty}^{\mu} e^{(x-\mu)/\sigma}\, dx + \frac{1}{2\sigma} \int_{\mu}^{\infty} e^{-(x-\mu)/\sigma}\, dx = 1.$$

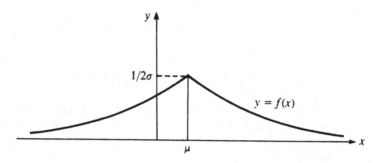

Laplace distribution

(d) $$\int_{-\infty}^{\infty} f(x)\, dx = \frac{1}{4} \int_{0}^{\infty} xe^{-x/2}\, dx = 1.$$

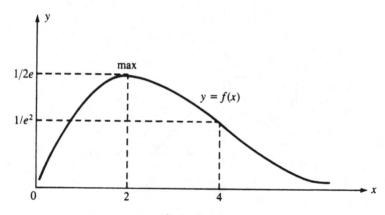

$\chi_2^{(2)}$-distribution

91. (i) We have

$$\int_{-\infty}^{\infty} f(x)\, dx = a \int_{0}^{3} x\, dx + a \int_{3}^{6} (6 - x)\, dx = 9a,$$

and the condition

$$\int_{-\infty}^{\infty} f(x)\, dx = 1$$

gives

$$a = \frac{1}{9}.$$

(ii)
$$P[X > 3] = \frac{1}{9} \int_3^6 (6 - x) \, dx = \frac{1}{2},$$
$$P[1.5 < X < 4.5] = \frac{1}{9} \int_{1.5}^3 x \, dx + \frac{1}{9} \int_3^{4.5} (6 - x) \, dx = \frac{3}{4}.$$

(iii)
$$P(AB) = P[3 < X < 4.5] = 3/8 = P(A) \cdot P(B).$$

Hence A and B are stochastically independent.

92. (a)
$$P[X > 5] = \frac{1}{5} \int_5^\infty e^{-x/5} \, dx = e^{-1}.$$

(b)
$$P[3 < x < 6] = \frac{1}{5} \int_3^6 e^{-x/5} \, dx = e^{-3/5} - e^{-6/5}.$$

(c)
$$P[X < 3] = \frac{1}{5} \int_0^3 e^{-x/5} \, dx = 1 - e^{-3/5}.$$

(d)
$$P[X < 6 | X > 3] = \frac{P[3 < X < 6]}{P[X > 3]} = 1 - e^{-3/5}.$$

93. (a)
$$P[0.1 < X < 0.2] = 0.1.$$

(b)
$$\sum_{K=0}^9 P[0.k5 < X < 0.k6] = 10 \times 0.01 = 0.1.$$

(c)
$$P[0.3 \le \sqrt{X} < 0.4] = P[0.09 \le X < 0.16] = 0.07.$$

94. (I) Let X be the height of a man in centimeters. Then X is $N(167.9)$.

(a)
$$P[X > 167] = P\left[\frac{X - 167}{3} > 0 \right] = P[Z > 0] = \frac{1}{2}$$
$$(Z \sim N(0, 1)).$$

(b)
$$P[X > 170] = P\left[\frac{X - 167}{3} > \frac{170 - 167}{3} \right]$$
$$= P[Z > 1] = 1 - \Phi(1) = 16\%,$$

where Φ denotes the cumulative distribution function of the standard normal $N(0, 1)$.

(II) (i) Let Y be the number of men that have height greater than 170 cm. Then Y obeys a binomial law with $n = 4$ and $p = P[X > 170] = 0.16$. Hence

$$P(Y = 4) = (0.16)^4 = 0.0007.$$

(ii) If Z denotes the number of men that have height greater than the mean $\mu = 167$, then the random variable Z has the binomial distribution with parameters $n = 4$ and $p = P[X > 167] = 0.5$. Thus

$$P[Z = 2] = \binom{4}{2} (0.5)^4 = 0.375.$$

95. (a) Let X be the length of a bolt. Then X is $N(5, (0.2)^2)$,

$$p = P[X \notin (4.8, 5.2)] = 1 - P[4.8 < X < 5.2]$$
$$= 1 - P[-1 < Z < 1] = 2(1 - \Phi(1)), \, p = 0.32.$$

(b) If Y denotes the number of defective bolts in the sample, then Y has the binomial distribution with $n = 10$, $p = 0.32$. Hence $P[Y = 0] = (0.68)^{10}$.

96. Let T be the time interval (in minutes) between two successive arrivals. Then T has the exponential distribution with mean 3, i.e.,

$$f(t) = \tfrac{1}{3}e^{-t/3}, \qquad t > 0.$$

(a) $P[T < 3] = 1 - e^{-1} = 0.37.$
(b) $P[T > 4] = e^{-4/3}.$
(c) Let X be the number of customers per hour and Y be the number of customers who buy the object. Then X has a Poisson distribution with $\lambda = 20$. On the other hand,

$$P[Y = k/X = n] = \binom{n}{k}(0.1)^k(0.9)^{n-k}, \qquad k = 0, 1, \ldots, n.$$

Hence

$$P[Y = k] = \sum_{n=k}^{\infty} P[Y = k|X = n]P[X = n]$$

$$= \sum_{n=k}^{\infty} \binom{n}{k}(0.1)^k(0.9)^{n-k}e^{-20}\frac{20^n}{n!} = e^{-2}\frac{2^k}{k!}, \qquad k = 0, 1, \ldots,$$

that is, Y has a Poisson distribution with parameter $\lambda = 2$.

97. We have

$$F_Y(y) = P[Y \leq y] = P[X^2 \leq y] = P[-\sqrt{y} < X < \sqrt{y}]$$
$$= F(\sqrt{y}) - F(-\sqrt{y}).$$

Since F is differentiable so is F_Y and Y is a continuous random variable with density

$$f_Y(y) = \frac{d}{dy}F_Y(y) = \frac{1}{2\sqrt{y}}[f(\sqrt{y}) + f(-\sqrt{y})]. \qquad (*)$$

98. Using $(*)$ of the preceding exercise, we obtain:

(a) $$f_Y(y) = \frac{1}{2\sigma\sqrt{2\pi y}}\exp\left[-\frac{y+\mu^2}{\sigma^2}\right], \qquad y > 0.$$

(b) $$f_Y(y) = \frac{1}{2\sqrt{y}}e^{-\sqrt{y}}, \qquad y > 0.$$

(c) $$f_Y(y) = \frac{1}{\pi}\frac{1}{\sqrt{y(1+y)}}, \qquad y > 0.$$

99. $F_Y(y) = P[|X| < y] = P[-y < X < y] = F(y) - F(-y),$

$$f_Y(y) = \frac{d}{dy} F_Y(y) = f(y) + f(-y), \qquad y > 0.$$

(a) $\qquad\qquad f_Y(y) = \frac{1}{\sigma\sqrt{2\pi}} \exp[-(y^2 + \mu^2)], \qquad y > 0.$

(b) $\qquad\qquad f_Y(y) = e^{-y}, \qquad y > 0.$

(c) $\qquad\qquad f_Y(y) = \frac{2}{\pi} \frac{1}{1 + y^2}, \qquad y > 0.$

100. $F_X(x) = P[X \le x] = P[\log X \le \log x] = P[Y \le \log x]$

$$= F_Y(\log x),$$

$$f_X(x) = \frac{d}{dx} F_X(x) = \frac{1}{x} f_Y(\log x)$$

$$= \frac{1}{\sigma\sqrt{2\pi}\, x} \exp\left[-\frac{(\log x - \mu)^2}{2\sigma^2} \right], \qquad x > 0.$$

101. $\qquad I \equiv \int_{-\infty}^{\infty} |x - m| f(x)\, dx = -\int_{-\infty}^{m} (x - m) f(x)\, dx$

$$+ \int_{m}^{\infty} (x - m) f(x)\, dx.$$

Applying Leibnitz's rule of differentiation,

$$\frac{d}{dy} \int_{p(y)}^{q(y)} f(x, y)\, dx = \int_{p(y)}^{q(y)} \frac{\partial}{\partial y} f(x, y)\, dx + f(q(y), y)q'(y) - f(p(y), y)p'(y),$$

we get

$$\frac{dI}{dm} = \int_{-\infty}^{m} f(x)\, dx - \int_{m}^{\infty} f(x)\, dx$$

and setting this equal to zero gives

$$F(m) = \int_{-\infty}^{m} f(x)\, dx = \frac{1}{2},$$

that is, m is the median of the distribution.

102. We have

$$E(X) = \int_{-\infty}^{\infty} x f(x)\, dx = \int_{-\infty}^{\infty} (x - a)f(x)\, dx + a \int_{-\infty}^{\infty} f(x)\, dx$$

$$= \int_{-\infty}^{a} (x - a)f(x)\, dx + \int_{a}^{\infty} (x - a)f(x)\, dx + a. \tag{1}$$

Since $E(X) < \infty$, the following integrals exist

$$\int_{-\infty}^{a} (x - a)f(x)\, dx = \int_{-\infty}^{0} yf(y + a)\, dy,$$

$$\int_{a}^{\infty} (x - a)f(x)\, dx = \int_{0}^{\infty} yf(y + a)\, dy,$$

and moreover, since $f(a + y) = f(a - y)$, we have

$$\int_{-\infty}^{0} yf(y + a)\, dy = -\int_{0}^{\infty} yf(a + y)\, dy.$$

Hence from (1) we obtain $E(X) = a$.

103.
$$E(tX + Y)^2 = E(t^2 X^2 + 2tXY + Y^2)$$
$$= t^2 E(X^2) + 2tE(XY) + E(Y^2).$$

Since $E(tX + Y)^2 \geq 0$ we get

$$t^2 E(X^2) + 2tE(XY) + E(Y^2) \geq 0 \quad \text{for every } t.$$

Therefore the (constant) coefficients of the quadratic in t must satisfy the relation

$$E^2(XY) - E(X^2)E(Y^2) \leq 0.$$

104.
$$|E(X)| = \left| \int_{-\infty}^{\infty} x\, dF(x) \right| \leq \int_{-\infty}^{\infty} |x|\, dF(x) = E(|X|).$$

105. We have

$$E[I_A(X)] = 1 \cdot P[X \in A] + 0 \cdot P[X \notin A] = P[X \in A] = P(A) = \int_A dF_X(x).$$

106. Applying Chebyshev's inequality, we get

$$P[|X - E(X)| > k\sigma] \leq \frac{1}{k^2},$$

since $E(X) = 0$ and $\Delta(X) = E(X^2) - [E(X)]^2 = 0$ it follows that

$$P[|X| > 0] \leq \frac{1}{k^2} \quad \text{for every positive integer } k.$$

Hence for every $\varepsilon > 0$ we have

$$P[|X| > 0] < \varepsilon.$$

This implies that $X = 0$ with probability 1.

107. $E(X - c)^2 = E(X - \mu + \mu - c)^2 = E(X - \mu)^2 + (\mu - c)^2,$

so that the minimum is attained when $c = \mu = E(X)$.

108. $\mu = E(X) = \int_{-\infty}^{\infty} x f(x) \, dx = \int_{-\infty}^{0} x \, dF(x) - \int_{0}^{\infty} x \, d[1 - F(x)]$

$= [xF(x)]_{-\infty}^{0} - [x(1 - F(x)]_{0}^{+\infty} - \int_{-\infty}^{0} F(x) \, dx + \int_{0}^{\infty} [1 - F(x)] \, dx$

$= \int_{0}^{\infty} [1 - F(x)] \, dx - \int_{-\infty}^{0} F(x) \, dx,$

where we used

$$\lim_{x \to -\infty} xF(x) = \lim_{x \to +\infty} x[1 - F(x)] = 0 \qquad \text{(Exercise 113)}.$$

109. $E(X - c)^k = \int_{-\infty}^{\infty} (x - c)^k f(x) \, dx$

$= \int_{-\infty}^{c} (x - c)^k \, dF(x) - \int_{0}^{\infty} (x - c)^k \, d[1 - F(x)]$

$= [(x - c)^k F(x)]_{-\infty}^{c} - [(x - c)^k (1 - F(x))]_{c}^{\infty}$

$\quad - \int_{-\infty}^{c} F(x) d(x - c)^k$

$\quad + \int_{c}^{\infty} [1 - F(x)] d(x - c)^k$

$= k \int_{c}^{\infty} (x - c)^{k-1} [1 - F(x)] \, dx$

$\quad - k \int_{-\infty}^{c} (x - c)^{k-1} F(x) \, dx,$

where we used

$$\lim_{x \to -\infty} (x - c)^k F(x) = \lim_{x \to +\infty} (x - c)^k [1 - F(x)] = 0.$$

110. The rth central moment is given by

$$\mu_r = E[(X - \mu)^r] = E\left[\sum_{k=0}^{r} (-1)^{r-k} \binom{r}{k} X^k \mu^{r-k} \right]$$

$$= \sum_{k=0}^{r} (-1)^{r-k} \binom{r}{k} E[X^k] \mu^{r-k}$$

$$= \sum_{k=0}^{r} (-1)^{r-k} \binom{r}{k} \mu_k' \mu^{r-k}, \qquad r = 1, 2, \dots, n.$$

The last relation shows that the first n moments determine the first n central moments. The inverse is also true because

$$\mu'_r = E[X^r] = E[(X - \mu) + \mu]^r = E\left[\sum_{k=0}^{r} \binom{r}{k}(X - \mu)^k \mu^{r-k}\right]$$

$$= \sum_{k=0}^{r} \binom{r}{k} \mu_k \mu^{r-k}$$

provided the mean μ is given.

111. Suppose X is continuous with density $f(x)$ (if X is discrete the proof is analogous). Then for every integer $r > 0$, we have

$$\int_{-\infty}^{\infty} |x|^r f(x) \, dx \leq \int_{-\infty}^{\infty} K^r f(x) \, dx = K^r < \infty,$$

hence $E(X^r)$ exists.

112. Since the nth order moment exists, the integral $\int_{-\infty}^{\infty} x^n f(x) \, dx$ converges absolutely, that is,

$$\int_{-\infty}^{\infty} |x|^n f(x) \, dx < +\infty.$$

For every $k = 1, 2, \ldots, n - 1$, we have

$$|x|^k < |x|^n + 1$$

and therefore

$$\int_{-\infty}^{\infty} |x|^k f(x) \, dx < \int_{-\infty}^{\infty} |x|^n f(x) \, dx + 1 < +\infty.$$

113. Since the mean of X exists, the integral $\int_{-\infty}^{\infty} |x| \, dF(x)$ also exists. Thus

$$0 = \lim_{x \to \infty} x[1 - F(x)] = \lim_{x \to \infty} x \int_{x}^{\infty} dF(y) \leq \lim_{x \to \infty} \int_{x}^{\infty} y \, dF(y) = 0$$

$$\Rightarrow \lim_{x \to -\infty} x[1 - F(x)] = 0.$$

Similarly, we can show that

$$\lim_{x \to -\infty} xF(x) = 0.$$

114. We have

$$V[S_n] = \sum_{k=1}^{n} V(X_k) + 2 \sum_{i<j} \text{Cov}(X_i, X_j) = n\sigma^2 + n(n - 1)\rho\sigma^2, \qquad (1)$$

where

$$V(X_k) = \sigma^2, \quad k = 1, 2, \ldots, n, \qquad \rho = \sigma^2 \text{Cov}(X_i, X_j), \quad i \neq j,$$

$$V(X_k) = E[X_k^2] - (E[X_k])^2.$$

But, since each of the numbers $1, 2, \ldots, N$ has probability $1/N$ of appearing in any selection, we have for $k = 1, 2, \ldots, n$

$$E(X_k) = \frac{1}{N} \sum_{k=1}^{N} k = \frac{N+1}{2} \tag{2}$$

and

$$E[X_k^2] = \frac{1}{N} \sum_{k=1}^{N} k^2 = \frac{(N+1)(2N+1)}{6}. \tag{3}$$

On the other hand, since

$$V\left[\sum_{k=1}^{N} X_k \right] = V(\text{constant}) = 0 = N\sigma^2 + N(N-1)\rho\sigma^2,$$

we get

$$\rho = -\frac{1}{N-1} \tag{4}$$

and from (1), (2), (3), and (4) we obtain

$$V[S_n] = \frac{n(n^2-1)}{12}\left[1 - \frac{n-1}{N-1} \right].$$

Note that if X_k $(k = 1, 2, \ldots, N)$ were independent, e.g., when we draw a sample from an infinite population (that is when $N \to \infty$) then we would have

$$V[S_n] = \frac{n(n^2-1)}{12}.$$

The dependence of X_k, when we draw a sample from a finite population, reduces the variance by the factor $100((n-1)/(N-1))\%$, referred to as *the finite population correction factor*.

115. Let X be the number of white balls in the sample. Then:

(a) X is binomial $b(k, n, p)$, where $p = N_1/(N_1 + N_2)$, and hence

$$E(X) = np = \frac{nN_1}{N_1 + N_2}.$$

(b) X is hypergeometric and

$$E(X) = \sum_k k \frac{\binom{N_1}{k}\binom{N_2}{n-k}}{\binom{N_1+N_2}{n}}.$$

This expectation is easily computed if we represent X as a sum

$$X = \sum_{r=1}^{n} X_r,$$

where

$$X_r = \begin{cases} 1 & \text{if the } r\text{th selection is a white ball,} \\ 0 & \text{if the } r\text{th selection is a black ball,} \end{cases} \quad r = 1, 2, \ldots, n,$$

are identically distributed and therefore

$$E(X_r) = P[\text{white ball at the } r \text{ selection}] = p = \frac{N_1}{N_1 + N_2}, \quad r = 1, 2, \ldots, n.$$

Hence

$$E(X) = \sum_{r=1}^{n} E(X_r) = np.$$

116. The random variable X takes on the values 0, 1, and 2 with probabilities

$$p_0 = P[X = 0] = \frac{2}{3} \cdot \frac{4}{5} = \frac{8}{15},$$

$$p_1 = P[X = 1] = \frac{1}{3} \cdot \frac{4}{5} + \frac{2}{3} \cdot \frac{1}{5} = \frac{6}{15},$$

$$p_2 = P[X = 2] = \frac{1}{3} \cdot \frac{1}{5} = \frac{1}{15},$$

respectively.

(a) The expected number of correct answers is

$$\mu = E(X) = 0 \cdot p_0 + 1 \cdot p_1 + 2 \cdot p_2 = \frac{6}{15} + 2 \cdot \frac{1}{15} = \frac{8}{15}.$$

(b)
$$E(X^2) = p_1 + 4p_2 = \frac{6}{15} + \frac{4}{15} = \frac{10}{15}.$$

Hence

$$V(X) = E(X^2) - [E(X)]^2 = \frac{10}{15} - \frac{64}{225} = \frac{86}{225}.$$

Another method. Let X_k be the number 0 or 1 of correct answers in problem k $(k = 1, 2)$. Then $X = X_1 + X_2$ and so

$$E(X) = E(X_1) + E(X_2) = P[X_1 = 1] + P[X_2 = 1] = \frac{1}{3} + \frac{1}{5} = \frac{8}{15},$$

$$V(X) = V(X_1) + V(X_2) = P[X_1 = 0]P[X_1 = 1] + P[X_2 = 0]P[X_2 = 1]$$

$$= \frac{2}{9} + \frac{4}{25} = \frac{86}{225}.$$

117. Let X be the gain from the ticket. Then X takes the values 1,000, 500, and 100 with probabilities 1/3,000, 1/3,000 and 5/3,000, respectively. Hence

$$E(X) = 1,000 \cdot \frac{1}{3,000} + 500 \cdot \frac{1}{3,000} + 100 \cdot \frac{5}{3,000} = \frac{2}{3}.$$

Since a man pays 1 dollar for the ticket the expected (total) gain is $2/3 - 1 = -1/3$, i.e., a loss.

118. (a) Let X be the number of throws. Then X obeys a Pascal (negative binomial) probability law with parameters $n = 3$ and $p = 2/3$. Hence the expected number of throws in a performance of the experiment is

$$E(X) = \frac{n}{p} = \frac{9}{2}.$$

(b) In 10 repetitions the expected number of throws is

$$10 \cdot E(X) = 45.$$

119. Let X be the total gain of the gambler. The random variable X takes the values $-1, 0, 1,$ and 7 with probabilities

$$P[X = -1] = P[\text{no ace}] = \left(\frac{5}{6}\right)^3 = \frac{125}{216},$$

$$P[X = 0] = P[1 \text{ ace}] = \binom{3}{1}\left(\frac{1}{6}\right)\left(\frac{5}{6}\right)^2 = \frac{75}{216},$$

$$P[X = 1] = P[2 \text{ aces}] = \binom{3}{2}\left(\frac{1}{6}\right)^2\left(\frac{5}{6}\right) = \frac{15}{216},$$

$$P[X = 2] = P[3 \text{ aces}] = \binom{3}{2}\left(\frac{1}{6}\right)^3 = \frac{1}{216}.$$

Hence

$$E(X) = (-1)\frac{125}{216} + 0 \cdot \frac{75}{216} + 1 \cdot \frac{15}{216} + 7 \cdot \frac{1}{216} = -\frac{103}{216},$$

which shows that the game is not fair. The game becomes fair if the gambler gets a dollars when three aces appear where a satisfies the relation

$$E(X) = -\frac{125}{216} + \frac{15}{216} + \frac{a-1}{216} = \frac{a-111}{216} = 0 \Rightarrow a = 111 \text{ dollars.}$$

120. Let p_n be the probability that the gambler obtains heads for the first time at the nth toss and X be the gambler's gain. Then X assumes the values -15 and 1 with probabilities

$$P[X = -15] = P[\text{all 4 tosses result in tails}] = \frac{1}{16},$$

$$P[X = 1] = p_1 + p_2 + p_3 + p_4 = \frac{1}{2} + \frac{1}{4} + \frac{1}{8} + \frac{1}{16} = \frac{15}{16}.$$

Hence $E(X) = 0$, that is, on the average, the gambler preserves his capital.

121. Let A_i be the event that book i arrives (reaches its destination), let A be the event that both books arrive, and let B be the event that at least one book arrives, and let μ denote the expected net value; then:
  (i) If each book is sent separately, we have:
    (a) $P(A) = 0.9 \times 0.9 = 0.81$.
    (b) $P(B) = 1 - P(\text{none reaches its destination}) = 1 - 0.1 \times 0.1 = 0.99$.
    (c) $\mu = 0 \times P(A_1^c A_2^c) + 2 \times P(A_1 A_2^c \cup A_1^c A_2) + 4 \times P(A_1 A_2) - 0.20$
         $= 0 \times 0.01 + 2 \times 0.18 + 4 \times 0.81 - 0.20 = 3.4$.
  (ii) If the books are sent in a single parcel, then:
    (a) $P(A) = 0.9$.
    (b) $P(B) = 0.9$.
    (c) $\mu = 0 \times 0.1 + 4 \times 0.9 - 0.15 = 3.45$.
Thus, according to criteria (a) and (c), a single parcel is preferable, whereas according to (b), method (i) is better.

122. (a) $X = 2X_0$ where X_0 is a Bernoulli random variable with parameter p. Hence

$$E(X) = 2E(X_0) = 2p, \qquad \text{Var}(X) = 4\,\text{Var}(X_0) = 4pq.$$

(b) $X = X_1 + X_2$ where X_1 and X_2 are independent Bernoulli with parameters p_1 and p_2, respectively. Hence

$$E(X) = E(X_1) + E(X_2) = p_1 + p_2, \qquad \text{Var}(X) = \text{Var}(X_1) + \text{Var}(X_2)$$

$$= p_1 q_1 + p_2 q_2.$$

123.

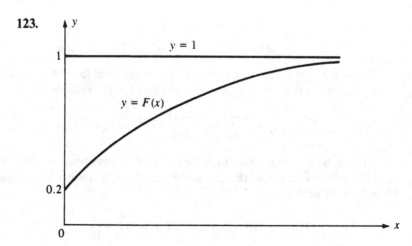

$$F(x) = \begin{cases} 1 - 0.8e^{-x}, & x \geq 0, \\ 0 & x < 0, \end{cases}$$

$$P[X = 0] = F(0) - F(0-) = 0.2,$$

$$f(x) = \begin{cases} 0.8e^{-x}, & x > 0, \\ 0, & x < 0, \end{cases}$$

$$E(X) = 0.8 \int_0^\infty x e^{-x} \, dx = 0.8.$$

Suppose that the lifetime of a certain kind of bulb obeys an exponential law with parameter $\lambda = 1$, and consider a box containing 80% of this kind of bulb and 20% defective bulbs. Then the lifetime of the bulbs in the box has the above distribution function $F(x)$.

124. $\quad E[Y] = \displaystyle\int_{-\infty}^\infty \frac{1 - \Phi(x)}{\varphi(x)} \varphi(x - \mu) \, dx$

$$= \frac{1}{\mu} \int_{-\infty}^\infty [1 - \Phi(x)] \exp\left[\frac{2\mu x - \mu^2}{2}\right] d\left[\frac{2\mu x - \mu^2}{2}\right]$$

$$= \frac{1}{\mu} [1 - \Phi(x)] \exp\left[\frac{2\mu x - \mu^2}{2}\right] \Big|_{-\infty}^{+\infty}$$

$$+ \frac{1}{\mu} \int_{-\infty}^\infty \exp\left[\frac{2\mu x - \mu^2}{2}\right] \varphi(x) \, dx$$

$$= 0 + \frac{1}{\mu} \int_{-\infty}^\infty \varphi(x - \mu) \, d(x - \mu) = \frac{1}{\mu}.$$

125. We have

$$E(X) = \frac{1}{e^\lambda - 1} \sum_{k=1}^\infty k \frac{\lambda^k}{k!} = \frac{\lambda}{e^\lambda - 1} \sum_{k=1}^\infty \frac{\lambda^{k-1}}{(k-1)!} = \frac{\lambda e^\lambda}{e^\lambda - 1} = \frac{\lambda}{1 - e^{-\lambda}},$$

$$E[X(X - 1)] = \frac{1}{e^\lambda - 1} \sum_{k=2}^\infty k(k-1) \frac{\lambda^k}{k!} = \frac{\lambda^2}{e^\lambda - 1} \sum_{k=2}^\infty \frac{\lambda^{k-2}}{(k-2)!} = \frac{\lambda^2}{1 - e^{-\lambda}},$$

and so

$$V(X) = E[X(X - 1)] + E(X) - [E(X)]^2 = \frac{\lambda^2 + \lambda}{1 - e^{-\lambda}} - \frac{\lambda^2}{(1 - e^{-\lambda})^2}$$

$$= \frac{\lambda(1 - \lambda e^{-\lambda} - e^{-\lambda})}{(1 - e^{-\lambda})^2}.$$

126. The elements (simple events) are the following:

$$AA, \, ACC, \, ACBB, \, ACBACC, \, ACBAA, \, ACBACBB, \ldots,$$

$$BB, \, BCC, \, BCAA, \, BCABCC, \, BCABB, \, BCABCAA, \ldots,$$

and

$ACBACBACB...$ (infinite sequence), $BCABCABCA...$ (infinite sequence).

If we assume that the games are independent, we conclude that the probability of a simple event consisting of k games equals $1/2^k$, because the probability of each player winning a game is $1/2$. Hence

$$\sum_{k=2}^{\infty} 2\frac{1}{2^k} = \frac{1}{2} + \frac{1}{4} + \frac{1}{8} + \cdots = 1.$$

Thus the last two elements of the sample space have zero probability and therefore the game terminates with probability 1.

127. The sample space consists of the elements:

$$HH, HTT, HTHH, HTHTT, HTHTHH, \ldots,$$

$$TT, THH, THTT, THTHH, THTHTT, \ldots,$$

and

$$\left.\begin{array}{l} HTHTHTHT\ldots \\ THTHTHTH\ldots \end{array}\right\} \text{ up to infinity, with zero probability.}$$

The required probabilities are

$$P(A) = \sum_{k=2}^{7} 2\frac{1}{2^k} = 1 - \left(\frac{1}{2}\right)^6 = \frac{63}{64},$$

$$P(B) = \sum_{k=1}^{\infty} \left(\frac{1}{2}\right)^{2k} \cdot 2 = \frac{2}{3}.$$

128. The number of all possible outcomes are $36^6 = 6^{12}$. The number of favorable outcomes are $12!/2^6$ and hence the required probability is $12!/(2^6 \cdot 6^{12}) = 0.0034$.

129. Let A_k be the event that the player k has a complete suit. Then the required probability is $p = P\left(\bigcup_{k=1}^{4} A_k\right)$ and by the Poincaré formula we get

$$p = 4P[A_k] - 6P[A_iA_j] + 4P[A_iA_jA_k] - P[A_1A_2A_3A_4],$$

where

$$P[A_k] = \frac{4}{\binom{52}{13}}, \qquad p[A_iA_j] = \frac{(4)_2}{\binom{52}{13}\binom{39}{13}},$$

$$P[A_iA_jA_k] = P[A_1A_2A_3A_4] = \frac{4!}{\binom{52}{13}\binom{39}{13}\binom{26}{13}}.$$

130. Let the event A be that I win the bet. Then, denoting the result "tails" by T, we have

$$A = \text{HHH} \cup \text{HHT} \cup \text{THH}.$$

Tossing the coin C_1 first, that is, playing according to the series $C_1 C_2 C_1$, we have

$$P(A) = p_1 p_2 p_1 + p_1 p_2 (1 - p_1) + (1 - p_1) p_2 p_1 = p_1 p_2 (2 - p_1).$$

On the other hand, tossing the coin C_2 first we have

$$P(A) = p_2 p_1 p_2 + p_2 p_1 (1 - p_2) + (1 - p_2) p_1 p_2 = p_1 p_2 (2 - p_2),$$

and, since $p_1 > p_2$,

$$p_1 p_2 (2 - p_1) < p_1 p_2 (2 - p_2),$$

that is, you must select coin C_2 for the first toss.

Remark. If one has to play against two persons, he should play first against the stronger player in order to maximize his chances of winning.

131. *Method 1.* If we denote by A_k the appearance of the kth pair, then the required probability is

$$p = P\left[\bigcup_{k=1}^{10} A_k \right].$$

Applying Poincaré's theorem, since $P(A_{i_1} A_{i_2} \ldots A_{i_k}) = 0$ for $k \geq 3$, we get

$$p = P\left[\bigcup_{k=1}^{10} A_k \right] = \sum_{k=1}^{10} P[A_k] - \sum_{i<j} P(A_i A_j) = 10 P(A_k) - \binom{10}{2} P(A_i A_j)$$

$$= 10 \frac{\binom{18}{2}}{\binom{20}{4}} - \binom{10}{2} \frac{1}{\binom{20}{4}} = \frac{99}{323}.$$

Method 2.

$$p = 1 - P(\text{no one pair}) = 1 - \frac{20 \cdot 18 \cdot 16 \cdot 14}{20 \cdot 19 \cdot 18 \cdot 17} = 1 - \frac{224}{323} = \frac{99}{323}$$

or

$$p = 1 - P(\text{selection of shoes belonging to 4 pairs}) = 1 - \frac{\binom{10}{4} \cdot 2^4}{\binom{20}{4}} = \frac{99}{323}.$$

For a random selection of 10 shoes the probability of selecting no pair is less than 6%.

132. (a) $(1/N)(1 - 1/N)^{n-1}$ (geometric with $p = 1/N$).

(b) $(1 - 1/N)^n$ because the first n balls must fall in the cells $2, 3, \ldots, N$.

133. (a) The n ($n \le N + 1$) first balls must fall in different cells. Thus the required probability is $(N)_n/N^n$.

(b) The probability, say p_n, that n throws will be necessary equals the probability that the first $n - 1$ balls will fall in different cells and the nth ball in one of the $n - 1$ occupied cells. Hence

$$p_n = \frac{(N)_{n-1}}{N^{n-1}} \frac{n - 1}{N},$$

and the expected number of throws is given by the sum

$$\sum_{n=i}^{N+1} n p_n = \sum_{n=i}^{N} \frac{n(n - 1)(N)_{n-1}}{N^n}.$$

134. (a) Let E_1 be the event that the nth card is the first ace. Then

$$P(E_1) = \frac{(48)_{n-1}(4)_1}{(52)_n}.$$

(b) Let E_2 be the event that the first ace appears among the first n cards

$$P(E_2) = 1 - \frac{(48)_n}{(52)_n}.$$

135. The thirteenth diamond may appear at the nth position ($26 \le n \le 52$). Of the 52! cases (permutations of the 52 cards), the number of favorable cases is equal to

$$\sum_{n=26}^{52} (n - 1)!(52 - n)! \binom{26}{n - 26} \cdot 13.$$

136. To each triplet of numbers out of the $\{1, 2, \ldots, n\}$ there correspond six permutations (ordered triplets), and of these, three have the first number smaller than the second. Since the $(n)_3$ ordered triplets are equiprobable, the required probability is $1/2$.

This also follows from the fact that the first number is equally likely to be larger or smaller than the second one.

137. Let $p(n, m)$ denote the probability that in placing n balls in m cells all cells are occupied. Then the required probability p_n is given by

$$p_n = p(n, m) - p(n - 1, m).$$

To find $p(n, m)$, let A_j be the event that the jth cell is empty. By the Poincaré formula (1.1)

$$p(n, m) = 1 - P(A_1 \cup \cdots \cup A_m) = 1 - S_1 + S_2 - \cdots + (-1)^m S_m$$

$$= \sum_{j=0}^{m} (-1)^j \binom{m}{j} \left(1 - \frac{j}{m}\right)^n,$$

since

$$S_k = \sum_{1 \le i_1 < \cdots < i_k \le n} P(A_{i_1} \ldots A_{i_k}) = \binom{m}{k} P(A_1 \ldots A_k) = \binom{m}{k} \left(1 - \frac{k}{m}\right)^n.$$

138. (a) The probability function of X is

$$P[X = k] = \frac{e^{-\lambda} \dfrac{\lambda^k}{k!}}{\displaystyle\sum_{k=0}^{n} e^{-\lambda} \dfrac{\lambda^k}{k!}} = \frac{1}{c} \frac{\lambda^k}{k!}, \qquad k = 0, 1, \ldots, n,$$

where we set

$$c = \sum_{k=0}^{n} \frac{\lambda^k}{k!}.$$

For the random variable Y we have

$$P[Y = ka - (n-k)b] = P[Y = k(a+b) - nb] = P[X = k]$$

$$= \frac{1}{c} \frac{\lambda^k}{k!}, \qquad k = 0, 1, \ldots, n.$$

Hence we find

$$E(Y) = \lambda(a+b) - \frac{(a+b)\lambda^{n+1}}{n! \, c} - nb.$$

(b) To maximize $E(Y)$, it suffices to minimize, with respect to n, the expression

$$\frac{(a+b)\lambda^{n+1}}{n! \, c} + nb.$$

139. Of the six permutations (cf. Exercise 136) only one is such that its first number is the smallest and the second number the largest. Hence the required probability is 1/6.

140. By the Poincaré formula (1.1) we have

$$P(AB \cup AC \cup BC) = P(AB) + P(AC) + P(BC) - 2P(ABC) = 0.104$$

using also the independence of A, B, C.

141. (a) Let X_n denote the outcome of the nth coin on the first throw and Y_n the outcome of the second throw ($n = 1, 2, 3, 4$). Then the required probability is

$$P[X_n = Y_n, n = 1, 2, 3, 4] = \prod_{i=1}^{4} P[X_n = Y_n] = \left(\frac{1}{2}\right)^4.$$

(b) Let X be the number of heads on the first throw and Y be the number of heads on the second throw; then

$$P[X = Y] = \sum_{k=0}^{4} P[X = k, Y = k] = \sum_{k=0}^{4} P[X = k]P[Y = k]$$

$$= \sum_{k=0}^{4} \binom{4}{k}^2 \frac{1}{2^8} = \frac{35}{128}$$

in virtue also of Exercise 21.

142. (a) $P[X = j] = \dfrac{(b)_{j-1}w}{(n)_j}, \qquad j = 1, 2, \ldots, b + 1.$

(b) $E[X] = \displaystyle\sum_{j=1}^{b+1} jP[X = j] = \lambda \sum_{j=1}^{\mu+1} \dfrac{j(b)_{j-1}}{(n)_j}.$

(c) The required relation follows from

$$\sum_{j=1}^{b+1} P[X = j] = \lambda \sum_{j=1}^{\mu+1} \frac{(b)_{j-1}}{(n)_j} = 1.$$

143. Obviously, when this occurs $2j$ balls will have been drawn ($j = 1, 2, \ldots, N$), and according to the hypergeometric distribution the required probability is equal to

$$\sum_{j=1}^{N} \binom{N}{j}\binom{N}{j} \Big/ \binom{2N}{2j}.$$

144. $P(A|C) = \dfrac{P(AC)}{P(C)} = \dfrac{P\left[\bigcup\limits_{j=1}^{\infty} ACB_j\right]}{P(C)} = \dfrac{\sum\limits_{j=1}^{\infty} P(ACB_j)}{P(C)}$

$$= \frac{\sum\limits_{j=1}^{\infty} P(A|CB_j)P(B_jC)}{P(C)} = \sum_{j=1}^{\infty} P(A|B_jC)P(B_j|C).$$

145. The probability of success for Rena is $1/3$ and the probability of failure is $2/3$. The probability of success for Nike is: $2/3 \cdot 1/2 = 1/3$. The probability of success for Galatea is: $2/3 \cdot 1/2 \cdot 1 = 1/3$. Therefore the process is fair (cf. random sampling from a finite population).

In the second case, the probability of success for Rena is a (say) and the probability for Nike is $b = a(2/3)$ and for Galatea is $c = a(2/3) \cdot (2/3)$. Hence

$$a = \frac{9}{19}, \quad b = \frac{6}{19}, \quad c = \frac{4}{19} \quad \text{since} \quad a + b + c = 1.$$

Therefore Nike and Galatea rightly protest.

146. Let A_i be the event that the ith face of the die appears. Let the given sides be $i_1, i_2, \ldots, i_k$. Then the required probability is

$$p_k = 1 - P\left[\bigcup_{n=1}^{k} A'_{i_n}\right] = 1 - \sum_{n=1}^{k} (-1)^{n-1} S_n,$$

where

$$S_1 = \sum_{n=1}^{k} P[A'_{i_n}] = \binom{k}{1}\left(\frac{5}{6}\right)^v,$$

$$S_2 = \sum_{n,m} P[A'_{i_n} A'_{i_m}] = \binom{k}{2}\left(\frac{4}{6}\right)^v,$$

....

$$S_n = \binom{k}{n}\left(\frac{6-n}{6}\right)^v.$$

Hence

$$p_k = 1 + \sum_{n=1}^{k} (-1)^n \binom{k}{n}\left(\frac{6-n}{6}\right)^v = \sum_{n=0}^{k} (-1)^n \binom{k}{n}\left(\frac{6-n}{6}\right)^v. \qquad (1)$$

Application. What is the probability that each face will appear at least once? Applying formula (1) for $k = 6$ we find

$$p_6 = \sum_{n=0}^{6} (-1)^n \binom{6}{n}\left(\frac{6-n}{6}\right)^v = \sum_{n=0}^{5} (-1)^n \binom{6}{n}\left(\frac{6-n}{6}\right)^n$$

(cf. Exercise 137).

147. From

$$\int_0^\infty f(x)\, dx = c \int_0^\infty \frac{1}{\sqrt{2\pi}} e^{-x^2/2}\, dx = \frac{c}{2} = 1$$

we have $c = 2$.

$$E(X) = \int_0^\infty x f(x)\, dx = \int_0^\infty 2x \frac{1}{\sqrt{2\pi}} e^{-x^2/2}\, dx = -\frac{2}{\sqrt{2\pi}} \int_0^\infty d(e^{-x^2/2})$$

$$= -\frac{2}{\sqrt{2\pi}} e^{-x^2/2}\Big|_0^\infty = \frac{2}{\sqrt{2\pi}} = \frac{\sqrt{2}}{\sqrt{\pi}}.$$

148. (a) Let X denote the lifetime in hours of an electric bulb. Then

$$q = P[X < 200] = P\left[\frac{X - 180}{20} < 1\right] = \Phi(1) = 0.84.$$

Therefore the required probability is

$$(1 - q)^4 = p^4 = (0.16)^4.$$

(b) Consider the event A_k where the urn contains k bulbs with lifetime greater than 200 hours, and the event B where the selected bulb has a lifetime greater than 200 hours.

$$P(B) = \sum_{k=1}^{4} P[(B|A_k)]P(A_k) = \sum_{k=1}^{4} \frac{k}{4}\binom{4}{k}p^k q^{4-k}$$

$$= p \sum_{k=1}^{4} \binom{3}{k-1}p^{k-1}q^{4-k} = p;$$

that is, the probability of selecting a bulb with lifetime greater than 200 hours is the same whether a bulb is chosen from the population of bulbs or from a random sample of size n (here $n = 4$) of the population.

149. Let X denote the outcome (sum) in a throw $2 \le X \le 12$. Then

$$P[X = 2] = P[X = 12] = 1/36, \qquad P[X = 3] = P[X = 11] = 1/18,$$

$$P[X = 4] = P[X = 10] = 1/12, \qquad P[X = 5] = P[X = 9] = 1/9,$$

$$P[X = 6] = P[X = 8] = 5/36, \qquad P[X = 7] = 1/6.$$

Let X_n denote the outcome of the nth throw, let W be the event the gambler (finally) wins, let W_1 be the event he wins on the first throw, and let L_1 be the event he loses on the first throw, then

$$P[W_1] = P[X = 7] + P[X = 11] = 1/6 + 1/18 = 2/9,$$

$$P[L_1] = P[X_1 = 2] + P[X_1 = 3] + P[X_1 = 12] = 1/9,$$

and therefore

$$P(\text{more than one throw}) = 1 - P(W_1) - P(L_1) = 2/3.$$

We now have for case (a)

$$P[W] = \frac{2}{9} + \sum_{\substack{j=4 \\ j \neq 7}}^{10} P[W|X_1 = j]P[X_1 = j], \tag{1}$$

and from

$$p_j = P[W|X_1 = j] = \sum_{v=2}^{\infty} P[X_v = 7]P[X_i \neq 7, X_i \neq j, i = 2, \ldots, v-1]$$

we find

$$p_4 = \sum_{r=0}^{\infty} \frac{1}{6}\left(1 - \frac{1}{6} - \frac{1}{12}\right)^r = \sum_{r=0}^{\infty} \frac{1}{6}\left(\frac{3}{4}\right)^r = \frac{2}{3} = p_{10},$$

$$p_5 = \sum_{r=0}^{\infty} \frac{1}{6}\left(1 - \frac{1}{6} - \frac{1}{9}\right)^r = \sum_{r=0}^{\infty} \frac{1}{6}\left(\frac{13}{18}\right)^r = \frac{3}{5} = p_9,$$

$$p_6 = \sum_{r=0}^{\infty} \frac{1}{6}\left(\frac{25}{36}\right)^r = \frac{6}{11} = p_8.$$

Substituting in (1) we find

$$P[W] = \frac{34}{55} = 0.618.$$

Under (b) we similarly obtain

$$P[W] = \frac{2}{9} + \sum_{\substack{j=4 \\ j \neq 7}}^{10} q_j P[X_1 = j] = \frac{244}{495} \approx 0.493,$$

where we set $p_j = 1 - q_j$. We notice that (b) defines an almost fair game. In fact, this is the most usual way of playing the game.

150. The expected profit for B is by Exercise 149

$$2 \cdot \frac{1}{36} + 3 \cdot \frac{2}{36} + 4 \cdot \frac{3}{36} + 5 \cdot \frac{4}{36} + 6 \cdot \frac{5}{36} + 7 \cdot \frac{6}{36} + 8 \cdot \frac{5}{36} + 9 \cdot \frac{4}{36} = \frac{188}{36}.$$

For A this is

$$\frac{3x}{36} + \frac{2x}{36} + \frac{x}{36} = \frac{6x}{36}.$$

The game is fair if $188/36 = 6x/36$. Hence $x = 31\frac{1}{3}$.

151. Let E_i be the event that one white ball and one red ball are drawn on the ith draw. Then by the multiplication rule the required probability is

$$P[E_1 E_2 \ldots E_n] = P[E_1] P[E_2 | E_1] \ldots P[E_n | E_1 \ldots E_{n-1}]$$

$$= \frac{n^2}{\binom{2n}{2}} \cdot \frac{(n-1)^2}{\binom{2n-2}{2}} \cdots \frac{1^2}{\binom{2}{2}} = \frac{(n!)^2}{\frac{(2n)!}{2^n}} = \frac{2^n (n!)^2}{(2n)!}.$$

Another method. The number of ways in which $2n$ balls may be divided into n pairs is $(2n)!/2^n$ of which $(n!)^2$ are favorable (for pairs of different color) because to every permutation of n white balls there correspond $n!$ permutations of red balls which give pairs of balls of different color.

152. Let E_i be the event that the number of the ball drawn from urn A is i and from B larger than i. Then the required probability is

$$p = P[E_1 \cup E_2 \cup \cdots \cup E_{n-1}] = \sum_{i=1}^{n-1} P(E_i) = \sum_{i=1}^{n-1} \frac{n-i}{n^2} = \frac{n-1}{2n}.$$

Another method. It is equally likely that the ball drawn from A will bear a number larger or smaller than that of B and the probability of equal numbers is $1/n$. Therefore

$$p + p + \frac{1}{n} = 1 \quad \text{and} \quad p = \frac{n-1}{2n}.$$

153. Let p_0 be the required probability. Then:

(i) (a) Every person after the second person chooses one of the other persons, except the first one. Therefore we find

$$p_0 = \left[\frac{(N-1)}{N}\right]^{n-1} = \left(1 - \frac{1}{N}\right)^{n-1}.$$

(b) Each of the n persons chooses one of those who have not been informed. Hence $p_0 = (N)_n/N^n$.

Similarly, when it is told to k persons at a time, we have

$$\text{(a)} \quad p_0 = \frac{(N-k)^{n-1}}{N^{n-1}} = \left[1 - \frac{k}{N}\right]^{n-1}, \qquad \text{(b)} \quad p_0 = \frac{(N)_{nk}}{[(N)_k]^n}.$$

(ii)

$$\text{(a)} \quad n = pN \quad \text{and} \quad \lim_{N\to\infty} p_0 = \lim_{N\to\infty}\left(1 - \frac{1}{N}\right)^{n-1} = e^{-p}.$$

$$\text{(b)} \quad \lim_{N\to\infty} p_0 = \lim_{N\to\infty}\left(1 - \frac{1}{N}\right)\left(1 - \frac{2}{N}\right)\cdots\left(1 - \frac{n-1}{N}\right).$$

154. Each gambler may win the game on the kth throw of the game (event A_k) where $k = n, n+1, \ldots, 2n-1$ with probability $\binom{k-1}{n-1}\left(\frac{1}{2}\right)^k$ because there are $\binom{k-1}{n-1}$ ways of winning $n-1$ throws out of $k-1$ throws and the kth throw must be a success.

Thus the probability that each player wins the game is

$$\sum_{k=n}^{2n-1} \binom{k-1}{n-1}\left(\frac{1}{2}\right)^k = \frac{1}{2}.$$

The probabilities after interruption are

$$P(I) = \sum_{v=n+k}^{2n-1} \binom{v-k-m-1}{n-k-1}\left(\frac{1}{2}\right)^{v-k-m},$$

$$P(II) = \sum_{v=n+k}^{2n-1} \binom{v-m-k-1}{n-m-1}\left(\frac{1}{2}\right)^{v-k-m}.$$

The amount will be divided in parts proportional to $P(I)$, $P(II)$.

155. Let A/B be a rational fraction and a an integer. The possible remainders of the division of A by a are $0, 1, \ldots, a-1$. Hence the probability that A is divisible by a is $1/a$. Similarly for B. Therefore the probability that both A and B are divisible by a is $1/a^2$. The fraction A/B is an irreducible fraction if and only if both A and B are not divisible by any of the prime numbers $2, 3, 5, \ldots$. Hence the required probability is given by

$$p = \left(1 - \frac{1}{2^2}\right)\left(1 - \frac{1}{3^2}\right)\left(1 - \frac{1}{5^2}\right)\cdots = \frac{6}{\pi^2}.$$

This is obtained by noting that

$$\frac{1}{p} = \sum_{a,b,c,\dots} \frac{1}{(2^a \cdot 3^b \cdot 5^c \cdots)^2} = \sum_{n=1}^{\infty} \frac{1}{n^2} = \frac{\pi^2}{6},$$

where the summation extends over all nonnegative integers a, b, c, ..., and taking into account that any positive integer n is of the form $n = 2^a \cdot 3^b \cdot 5^c \cdots$.

156. Let A_p be the event that the chosen number is divisible by the prime p. If p_1, p_2, ... are different prime divisors of n the events A_{p_1}, A_{p_2}, ... are independent. The probability that the chosen number is prime with respect to n is, by definition of $\varphi(n)$, $\varphi(n)/n$.

On the other hand, $P(A_p) = 1/p$, because there are q numbers out of 1, 2, 3, ..., n which are divisible by p, where $n = pq$. Thus

$$P(A_p) = \frac{q}{n} = \frac{q}{p \cdot q} = \frac{1}{p};$$

therefore

$$\frac{\varphi(n)}{n} = P\left[\bigcap_{p|n} A'_p\right] = \prod_{p|n} P(A'_p) = \prod_{p|n} \left(1 - \frac{1}{p}\right).$$

157. The generating function of the number of insects

$$S_N = \sum_{j=1}^{N} Y_j,$$

where N is the number of colonies and Y_j is the number of insects in the jth colony, is given by

$$P_{S_N}(t) = \exp\{\lambda(P(t) - 1)\},$$

where $P(t)$ is the generating function of Y_j. But

$$P(t) = \frac{\log(1 - pt)}{\log(1 - p)}$$

and therefore

$$P_{S_N}(t) = \exp\left\{\lambda\left[\frac{\log(1 - pt)}{\log(1 - p)} - 1\right]\right\},$$

that is, the probability generating function of the negative binomial distribution.

158. Let A_k be the event that the lot contains k defective tubes, and let E_r be the event that in a sample of n tubes r are defective. Then

(a)

$$P[E_r|A_k] = \binom{N-k}{n-r}\binom{k}{r} / \binom{N}{n},$$

$$P[A_k|E_r] = \frac{P[E_r|A_k]P[A_k]}{\sum_{j=r}^{m} P[E_r|A_j]P[A_j]} = \frac{p_k P[E_r|A_k]}{\sum p_j P[E_r|A_j]} \qquad \text{(Bayes's formula)}.$$

(b)

$$1 - \frac{\sum_{k=r}^{d-1} \binom{N-n}{k-r} \Big/ \binom{N}{k}}{\sum_{j=r}^{m} p_j \binom{N-n}{j-r} \Big/ \binom{N}{j}}.$$

159. Let A_k be the event where k defective articles are bought, let B_n be the event where n customers come into the shop, and let E be the event where a customer buys a defective article. Then we have

$$P(E) = P(\text{of buying})P(\text{buying a defective item}|\text{buying})$$

$$= \frac{2}{3} \cdot \frac{3}{4} = \frac{1}{2},$$

$$P(B_n) = p^n q, \qquad P[A_k|B_n] = \binom{n}{k}\frac{1}{2^n},$$

and

$$P[B_n|A_k] = \binom{n}{k}\frac{1}{2^n}qp^n \sum_{r=k}^{\infty} \binom{r}{k} 2^{-r}qp^r = \binom{n}{k} p^{n-k}(2-p)^{k+1}/2^{n+1}.$$

160. Let W denote the weight of an article. Then the proportion of articles with weight less than W_0 is

$$p = P[W < w_0] = \frac{1}{\sqrt{2\pi}} \int_{-\infty}^{w_0} e^{-(w-\mu)^2/2}\, dw = \frac{1}{\sqrt{2\pi}} \int_{-\infty}^{w_0-\mu} e^{-w^2/2}\, dw.$$

Then the expected profit is

$$K = (1-p) - (a + b\mu).$$

From

$$\frac{dK}{d\mu} = \frac{1}{\sqrt{2\pi}} e^{-(w_0-\mu)^2/2} - b = 0$$

we have the value of μ $(\mu > w_0)$ which maximizes the expected profit.

161. $P[\text{the clerk is late}] = P[\text{the train arrives after } 8:45] + P[\text{the train arrives before } 8:45]P[\text{the bus arrives after } 9] = 1 - \Phi(3/4) + \Phi(3/4)[1 - \Phi(2/3)] = 1 - \Phi(3/4)\Phi(2/3) \approx 0.4$.

(a) $P[\text{that the clerk is late}]P[\text{that the employer is late}] = 0.4 \times 0.07 = 0.028$ because we have $P[\text{that the employer is late}] = 1 - \Phi(3/2) = 0.07$.

(b) $P[\text{that the employer arrives before the clerk}] = P[\text{that the train arrives before } 8:45]P[\text{the bus arrives after the employer's car}] + P[\text{that the train arrives after } 8:45] = \Phi(3/4)\Phi(1/\sqrt{13}) + 1 - \Phi(3/4) \approx 0.5$ because the arrival time X of the bus obeys the normal probability law $N(8:58, 3^2)$ and the arrival time Y of the private car follows the $N(8:57, 2^2)$ and therefore

$$P[X > Y] = P[X - Y > 0] = P\left[\frac{(X - 8:58) - (Y - 8:57)}{\sigma_{X-Y}} > -\frac{1}{\sigma_{X-Y}}\right]$$

$$= \Phi\left(\frac{1}{\sqrt{13}}\right) \approx 0.61.$$

162. This is due to Dodge (see Rahman, 1967, p. 159).

(a) Let E_n be the event that the nth article examined is the first defective article $(n = 1, 2, \ldots, r)$, and let A be the event that the sequence is defective. Then

$$a = P[A] = P[E_1 \cup E_2 \cup \cdots \cup E_r] = \sum_{n=1}^{r} P(E_n) = \sum_{n=1}^{r} pq^{r-1} = 1 - q^r. \quad (1)$$

a is also the probability that among r articles at least one is defective. Hence (1).

(b) Let X be the number of examined articles of a defective sequence. Then

$$\lambda = E[X|A] = \sum_{k=1}^{r} \frac{kP[X = k]}{P[A]} = p \sum_{k=1}^{r} \frac{kq^{k-1}}{1 - q^r}$$

$$= \frac{1 - q^r(1 + rp)}{p(1 - q^r)}.$$

(i) Let Y be the number of defective sequences. Then Y obeys the geometric probability law with parameter $1 - P(A) = 1 - a$ (probability of success).

$$\mu = E(Y) = \sum_{n=0}^{\infty} n(1 - a)a^n = \frac{1 - q^r}{q^r} = q^{-r} - 1.$$

(ii)
$$\lambda\mu + r = \frac{1 - q^r}{pq^r} = \tau.$$

(c) Let φ denote the required percentage. Then

$\varphi = $ (average of examined articles before a defective article $+ \tau)/(\tau + \tau^*)$

$$= f/[f + q^r(1 - f)],$$

where we put τ^* as the average of articles which passed before a defective one was found; and $f^{-1} \times$ the average number of examined articles before a defective one was found.

(d)
$$\rho = p(1 - \varphi) = \frac{pq^r(1 - f)}{f + q^r(1 - f)}.$$

(e) ρ is maximized when $p = p^*$ as obtained from

$$(1 - f)(1 - p^{*r}) = \frac{f[(r + 1)p^* - 1]}{1 - p^*}.$$

For a more detailed discussion of Exercises 163–170 we refer to Mosteller (1965).

163. The error in A's syllogism is due to a mistaken sample space. He thinks each of the pairs of convicts AB, AC, BC, who are going to be set free, has probability $1/3$. This is correct from the point of view of the Board. But when we keep in mind the answer of the guard, we have the events:

1. A and B free and the guard answers B (B_1 event).

2. A and C free and the guard answers C (C_1 event).

3. B and C free and the guard answers B (B_2 event).

4. B and C free and the guard answers C (C_2 event).

Then

$$P[B_1] = P[C_1] = 1/3, \qquad P[B_2) = P[C_2] = 1/3 \cdot 1/2 = 1/6,$$

$$P[A \text{ free if the guard says } B] = \frac{P[A \text{ free and the guard says } B]}{P[\text{the guard says } B]}$$

$$= \frac{P(B_1)}{P(B_1) + P(B_2)} = \frac{1/3}{1/3 + 1/6} = 2/3.$$

Similarly, if the guard answers that C is set free.

Another method. It is equally likely that the guard is going to say that B (B event) or C (C event) is set free. Then, if A is the event that A is going to be set free, we have

$$P[A|B] = \frac{P[AB]}{P[B]} = \frac{1/3}{1/2} = \frac{2}{3}.$$

164. We obtain a number from the first box. The probability of a new number from the next box is $5/6$. Therefore, on average, $1/(5/6) = 6/5$ are required for the next new number because the mean of the geometric distribution is $1/p$.

Similarly, $6/4$ boxes will be required for the third number, $6/3$ for the fourth, $6/2$ for the fifth, and 6 for the sixth. Hence the required number of boxes on average is

$$6\left(\frac{1}{6} + \frac{1}{5} + \frac{1}{4} + \frac{1}{3} + \frac{1}{2} + 1\right) = 14.7.$$

For n coupons and large n, the required number is $n \log n + 0.577n + 1/2$ boxes because

$$1 + \frac{1}{2} + \cdots + \frac{1}{n} \approx \log n + \frac{1}{2n} + C,$$

where $C = 0.57721$ denotes Euler's constant.

165. The probability of a couple at the first two seats is

$$\frac{10}{19}\cdot\frac{9}{18} + \frac{9}{19}\cdot\frac{10}{18} = \frac{10}{19},$$

and the expected number of pairs at the first two seats is also

$$\frac{10}{19} = \frac{10}{19}\cdot 1 + \frac{9}{19}\cdot 0.$$

The same is true for every pair of successive seats and there are $19 - 1 = 18$ such pairs. Therefore the expected mean number of pairs (male–female student or female–male student) is $18\cdot 10/19 = 9\frac{9}{19} = 9.47$. More generally, for a male students and b female students the mean number of couples is

$$(a + b - 1)\left[\frac{ab}{(a+b)(a+b-1)} + \frac{ab}{(a+b-1)(a+b)}\right] = \frac{2ab}{a+b}.$$

Remark. We used the fact that the mean of a sum of random variables is equal to the sum of the mean values of the corresponding random variables, no matter whether the variables are (statistically) independent or not.

166. If A shoots and hits C, then B will certainly hit him; therefore he should not shoot at C. If A shoots at B and misses, then B will shoot at the more dangerous, i.e., C, and A has a chance to shoot at B and hit him with probability 0.4.

Naturally, if A fails then he has no chance. On the other hand, suppose that A hits B. Then C and A shoot alternately at each other until either one of them is hit. The probability that A will win is

$$(0.5)(0.4) + (0.5)^2(0.6)(0.4) + (0.5)^3(0.6)^2(0.3) + \cdots, \qquad (1)$$

where the nth term expresses the probability that in $2n$ trials C will fail n times, B will fail $n - 1$ times, and A will hit in the end. The sum in (1) is

$$\frac{(0.5)(0.4)}{1 - (0.5)(0.6)} = \frac{2}{7} = 0.28 < 0.4.$$

Thus if A hits B and continues with C, A has a smaller probability of winning than if the first shot fails. Hence to start, A shoots in the air and tries to hit B in the second round. Then C has no chance (of surviving).

167. The expected appearances of a 6 in the experiments of A, B, C are 1, 2, 3, respectively, but this does not mean that the corresponding probabilities of the events are equal. In fact, we have

$$P(A) = 1 - P(\text{no } 6) = 1 - \left(\frac{5}{6}\right)^6 \approx 0.665,$$

$$P(B) = 1 - P(\text{no } 6) - P(\text{one } 6) = 1 - \left(\frac{5}{6}\right)^{12} - \binom{12}{1}\left(\frac{1}{6}\right)\left(\frac{5}{6}\right)^{11} = 0.619,$$

$$P(C) = 1 - P(\text{no } 6) - P(\text{one } 6) - P(\text{two } 6\text{'s})$$

$$= 1 - \left(\frac{5}{6}\right)^{18} - \binom{18}{1}\left(\frac{1}{6}\right)\left(\frac{5}{6}\right)^{17} - \binom{18}{2}\left(\frac{1}{6}\right)^2\left(\frac{5}{6}\right)^{16} = 0.597.$$

Indeed Newton advised Pepys to choose the throw of six dice.

168. Employing one worker secures 364 working days; employing two workers, in all probability having different birthdays, we secure $2 \times (365 - 2) = 726$ working man-days, whereas if the number of workers is large there is a great probability that every day is a holiday because of somebody's birthday. Hence there is an optimum number of workers to be employed by the factory.

Consider n workers and let N be the number of days in a year. The probability that any day is a working day is equal to

$$p = P[\text{no birthday on a given day}] = [(N - 1)/N]^n = (1 - 1/N)^n,$$

and this day contributes, on average, np man-days. This is true of any day. Hence the total expected number of man-days during the whole period of N days when n workers are employed is equal to nNp. This is maximized for some $n = n_0$ if

$$(n_0 + 1)N\left(1 - \frac{1}{N}\right)^{n_0+1} \le n_0 N\left(1 - \frac{1}{N}\right)^{n_0},$$

and

$$n_0 N\left(1 - \frac{1}{N}\right)^{n_0} \ge (n_0 - 1)\left(1 - \frac{1}{N}\right)^{n_0-1},$$

from which we obtain

$$(n_0 + 1)\left(1 - \frac{1}{N}\right) \le n_0 \quad \text{and} \quad n_0 - 1 \le n_0\left(1 - \frac{1}{N}\right),$$

and finally

$$n_0 < N < n_0 + 1.$$

$n_0 = N - 1$ and $n_0 = N$ give the same maximum, i.e., $N^2(1 - (1/N))^N$; hence take $n_0 = N - 1$. For large N, $(1 - N^{-1})^N \approx e^{-1}$ and hence the expected number of man-days is $N^2 e^{-1}$, while if every day was a working day it would be N^2. Thus in a usual year (365 days) the $n_0 = 364$ workers will offer about 49,000 man-days. No doubt the wages must be very low in Erehwon.

169. There are several versions of the problem depending on the way in which a chord is chosen. We consider three versions.

(a) The middle of the chord is chosen at random, i.e., it is uniformly distributed over the circle. In this case, we have

$$P[(AB) < r] = P[M \notin C'] = 1 - \frac{\text{area } C'}{\text{area } C} = 1 - \frac{3}{4} = \frac{1}{4},$$

where C' is a circle with radius (see Fig. 1)

$$(KM) = r\cos(\pi/6) = r\sqrt{3}/2.$$

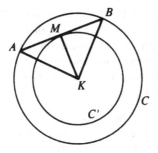

Figure 1

(b) Since A is some point on the circumference, take it fixed and choose B. Clearly now

$$P[(AB) < r] = \frac{120}{360} = \frac{1}{3},$$

since if B is chosen 60° on either side of A on the circumference, (AB) will be smaller than r.

(c) For reasons of symmetry, we suppose that AB has a given direction, say, of the diameter OO'; in this case, the middle M of the chord is uniformly distributed on the diameter KK' which is perpendicular to OO' (Fig. 2). We now have

$$P[(AB) < r] = P[(TM) > (TL)] = P\left[(TM) > \frac{\sqrt{3}}{2}r\right]$$

$$= 1 - \frac{2r\sqrt{3}/2}{2r} = 1 - \frac{\sqrt{3}}{2} = 0.13.$$

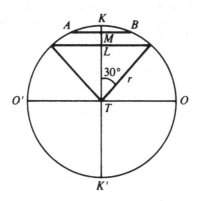

Figure 2

Remark. The variety of solutions exhibited is due to insufficient description of the corresponding random experiment which determines the choice of the chord. Thus in (a) we essentially choose at random a point M in the circle and require the probability that M lies in a smaller homocentric cycle. In (b) the points A and B are randomly chosen on the circumference, whereas in (c) we can imagine a cyliner rolling on a diameter of the circle (say KK') and the equally likely events correspond to equal intervals on the diameter (irrespective of their position on the diameter).

170. This is another version of the beauty contest problem or the dowry problem (how to select the most beautiful candidate or the highest dowry without being able to go back and choose a candidate already gone). The neophyte at a given stage of the race knows the rank (first, second, etc.) of the horse passing by; only the horse having the highest rank (up to that point) may be considered for final selection. Such horses will be called "candidates".

Before we consider the general case, let us see some simple cases. For $n = 2$ the neophyte has a chance of $1/2$ of choosing the better horse. For $n = 3$, choosing at random, e.g., the first horse he chooses has a probability $1/3$ of winning since of the six cases 123, 132*, 231*, 213*, 312, 321 only the last two are the favorable ones. The probability gets higher if he ignores the first horse and chooses the next better one. This procedure gives the three cases marked with an asterisk and therefore the probability becomes $1/2$.

In general, it will be shown that the best strategy is to let $s - 1$ horses pass and choose the first candidate after that. Let W_i denote the event that the ith horse wins. Then the ith horse will be chosen (as the best one) if

$$P[W_i] > P[W_{i+k}], \qquad k = 1, 2, \dots . \tag{1}$$

We will show that the right-hand side is a decreasing function of i, whereas the left-hand side is an increasing function of i, so that at some stage the process of observing more horses should stop and a decision taken. We observe that the probability that the ith horse is the best one is equal to the probability that the highest rank is among the first i ones, i.e., i/n, that is, an increasing function of i and at some stage it is larger than the probability of winning later. According to this strategy, the neophyte letting s horses pass has a probability of winning equal to the probability of coming across a single candidate after the nth stage. Thus we have

$P[W_k] = P[k$th horse has highest rank$]$

$\times P[$best of the first $k - 1$ horses is among the first $s - 1$ horses$]$

$$= \frac{1}{n} \frac{s - 1}{k - 1},$$

for $s \le k \le n$ and hence

$$\pi(s, n) = P[\text{to win}] = \sum_{k=s}^{n} P[W_k] = \frac{1}{n} \sum_{k=s}^{n} \frac{s - 1}{k - 1}$$

$$= \frac{s - 1}{n} \sum_{k=s-1}^{n-1} \frac{1}{k}, \qquad 1 < s \le n. \tag{2}$$

Certainly the first horse is a candidate and $\pi(1, n) = 1/n$. The optimum value s^* of s is the smallest integer for which (1) holds, i.e.,

$$\frac{s}{n} > \pi(s + 1, n) = \frac{s}{n}\left(\frac{1}{s} + \frac{1}{s + 1} + \cdots + \frac{1}{n - 1}\right)$$

or

$$\frac{1}{s} + \frac{1}{s + 1} + \cdots + \frac{1}{n - 1} < 1 < \frac{1}{s - 1} + \frac{1}{s} + \cdots + \frac{1}{n - 1}. \qquad (3)$$

For large n the approximation

$$1 + \frac{1}{2} + \cdots + \frac{1}{n} \approx \log n + C,$$

where $C = 0.57721$ is Euler's constant, gives

$$\pi(s, n) \approx \frac{s - 1}{n} \log \frac{n - 1}{s - 1} \approx \frac{s}{n} \log \frac{n}{s},$$

and by (3), $s^* \approx ne^{-1}$, i.e., for large n the neophyte must wait until $e^{-1} = 37\%$ of the horses pass by and then choose the first candidate. It can be found that for $n = 10$, $s^* = 4$, and $\pi(s, n) = \pi(4, 10) = 0.399$; similarly, $\pi(8.20) = 0.384$, $\pi(38,100) = 0.371$ thus approaching e^{-1} as $n \to \infty$. A random choice of a horse as the best would give $1/n$.

171. (a)

$$F_X(x) = 0, \quad x < 0, \qquad F_X(x) = 2/3, \quad 0 \le x < 1, \qquad F(x) = 1, \quad x \ge 1,$$

$$F_Y(y) = 0, \quad y < 0, \qquad F_Y(y) = 1/3, \quad 0 \le y < 1, \qquad F(y) = 1, \quad y \ge 1,$$

$$F(x, y) = \begin{cases} 0 & \text{for } x < 0 \text{ or } y < 0, \\ 1/3 & \text{for } (0 \le x < \infty, 0 \le y < 1), \\ 2/3 & \text{for } (0 \le x < 1, 1 \le y < \infty), \\ 1 & \text{for } x \ge 1, \quad y \ge 1. \end{cases}$$

(b) $(-1/2, 0)$ is a continuity point. $(0, 2)$ is a discontinuity point of $F(x, y)$ because for $x < 0$, $F(x, 2) = 0$, while $F(0, 2) = 2/3$.

172. $F(x, y)$ must satisfy $F(x_2, y_2) - F(x_1, y_2) - F(x_2, y_1) + F(x_1, y_1) \ge 0$ for all $x_1 \le y_1, x_2 \le y_2$ (why?); for $(x_1, y_1) = (0, 0), (x_2, y_2) = (2, 2)$ we have

$$F(2, 2) - F(2, 0) - F(0, 2) + F(0, 0) = 1 - 1 - 1 + 0 = -1;$$

thus it is not a distribution function.

173.
$$P[X_i = 1] = P[X_i = 0] = 1/2, \qquad i = 1, 2, 3;$$

$$P[X_i = a_i, X_j = a_j] = P[X_i = a_i]$$

for all $i \ne j$ $(i, j = 1, 2, 3)$ and $a_i, a_j = 0, 1$; then X_1, X_2, X_3 are pairwise independent. But the relation

$$P[X_1 = a_1, X_2 = a_2, X_3 = a_3] = P[X_1 = a_1]P[X_2 = a_2]P[X_3 = a_3]$$

does not hold, for example, for $a_1 = a_2 = a_3 = 1$; then

$$P[X_1 = 1, X_2 = 1, X_3 = 1] = 0 \neq \prod_{i=1}^{3} P[X_i = 1] = \frac{1}{8}.$$

174. Putting $x = yu$ we have,

$$f_Y(y) = ce^{-y} \int_0^y x^{n_1-1}(y - x)^{n_2-1}\, dx = ce^{-y}y^{n_1+n_2-1} \int_0^1 u^{n_1+n_2-1}(1 - u)^{n_2-1}\, du$$

$$= cy^{n_1+n_2-1}e^{-y} \frac{\Gamma(n_1)\Gamma(n_2)}{\Gamma(n_1 + n_2)},$$

and from the equation

$$\int_0^\infty f_Y(y)\, dy = c\Gamma(n_1)\Gamma(n_2).$$

we have $c^{-1} = \Gamma(n_1)\Gamma(n_2)$ and $F_Y(y)$ is a Gamma distribution with parameter $n_1 + n_2$. Similarly, we find that X has the Gamma distribution with parameter n_1:

$$f_X(x) = cx^{n_1-1} \int_x^\infty (y - x)^{n_2-1}e^{-y}\, dy = \frac{1}{\Gamma(n_1)} x^{n_1-1}e^{-x}, \qquad x > 0.$$

175. For $n_i \geq 0$ and $n_1 + \cdots + n_{k-1} + n_k \leq n$, we have

$$P[X_k = n_k | X_1 = n_1, \ldots, X_{k-1} = n_{k-1}]$$

$$= \frac{\binom{Np_1}{n_k}\binom{Np_{k+1}}{n_{k+1}} \Big/ \binom{N}{n}}{\binom{Np_{k-1}}{n_1} \cdots \binom{Np_{k-1}}{n_{k-1}}\binom{N(p_k + p_{k+1})}{n - x_k}} = \frac{\binom{Np_k}{n_k}\binom{Np_{k+1}}{n_{k+1}}}{\binom{N(p_k + p_{k+1})}{n - x_k}}$$

where $x_k = n - (x_1 + x_2 + \cdots + x_{k-1})$. This is hypergeometric.

176. Similarly as Exercise 175.

177. The marginal distribution of $X_1, \ldots, X_r$ $(r < k)$ is Dirichlet with density

$$f(x_1, \ldots, x_r) = \frac{\Gamma(n_1 + n_2 + \cdots + n_{k+1})s_{r+1}^{n_k+N_r-1}}{\Gamma(n_1)\ldots\Gamma(n_r)\Gamma(n_{r+1} + \cdots + n_{k+1})} \prod_{i=1}^{r} x_i^{n_i-1}, \qquad (1)$$

where we put

$$s_{r+1} = 1 - (x_1 + x_2 + \cdots + x_r), \qquad N_r = n_{r+1} + \cdots + n_{k+1}.$$

Equation (1) for $r = 1$ and $r = k - 1$ gives as density of X_k for given $X_i = x_i$
$(i = 1, 2, \ldots, k - 1)$,

$$f(x_k | X_1 = x_1, \ldots, X_k = x_{k-1}) = \frac{\Gamma(n_k + n_{k+1})}{\Gamma(n_k)\Gamma(n_{k+1})} \left(\frac{x_k}{s_k}\right)^{n_k - 1} \left(1 - \frac{x_k}{s_k}\right)^{n_{k+1} - 1} \frac{1}{s_k}.$$

Then the conditional distribution of X_k / R_k for $S_k = s_k$ is $\beta(n_k, n_{k+1})$.

178. The density is

$$f(x, y, z) = 1/V,$$

$$(x, y, z) \in S = \{(x, y, z), x \geq 0, y \geq 0, z \geq 0, x + y + z \leq c\},$$

(1)

where $V = $ volume of $(S) = c^3/6$. The joint density of (X, Y) is

$$f(x, y) = \frac{6}{c^3} \int_0^{c-x-y} dz = \frac{6(c - x - y)}{c}.$$

Note. For $c = 1$, (1) and (2) are special cases of Exercise 177.

179. (a) Since the quadratic form (see (5.10)) $x^2 - xy + y^2$ is positive definite, $f(x, y)$ is the density of the bivariate distribution $N(\mu, \Sigma)$ where

$$\mu = \begin{pmatrix} 0 \\ 0 \end{pmatrix}, \qquad \Sigma = \frac{1}{3} \begin{pmatrix} 2 & 1 \\ 1 & 2 \end{pmatrix}.$$

Hence

$$c = (2\pi\sqrt{|\Sigma|})^{-1} = \frac{\sqrt{3}}{2\pi}.$$

(b) The ellipses $x^2 - xy + y^2 = constant$.

180. (a) As in Exercise 178, we find

$$f(x, y) = 1/2, \qquad x \geq 0, \quad y \geq 0, \quad x + y \leq 2.$$

(b)

$$f_X(x) = \frac{1}{2} \int_0^{2-x} dy = \frac{1}{2}(2 - x).$$

(c)

$$f_Y(y | X = x) = \frac{f(x, y)}{f_X(x)} = \frac{1}{2 - x}, \qquad 0 < y < 2 - x.$$

181. (a) Multinomial $p_1 = p_2 = \cdots = p_6 = 1/6$, $n_1 = n_2 = \cdots = n_6 = 2$, $n = 12$.

(b)

$$P[X = i, Y = j] = \frac{12!}{i! \, j! \, (12 - i - j)!} \left(\frac{1}{6}\right)^{i+j} \left(\frac{2}{3}\right)^{12-i-j}, \qquad 0 \leq i + j \leq 12.$$

Trinomial with $p_1 = p_2 = 1/6$, $p_3 = 2/3$.

(c) $\operatorname{Cov}(X, Y) = E(XY) - E(X)E(Y)$,

$$E(X) = np_1 = 12 \cdot \frac{1}{6} = 2, \qquad E(Y) = np_2 = 12 \cdot \frac{1}{6} = 2,$$

$$E(XY) = \sum_{0 \le i+j \le n} ij \frac{p_1^i p_2^j p_3^{n-i-j}}{i! \, j! \, (n-i-j)!}$$

$$= n(n-1)p_1 p_2 \sum_{2 \le i+j \le n} \frac{(n-2)! \, p_1^{i-1} p_2^{j-1} p_3^{n-i-j}}{(i-1)! \, (j-1)! \, (n-i-j)!}$$

$$= n(n-1)p_1 p_2 \sum_{0 \le s+k \le m} \frac{m! \, p_1^s p_2^k p_3^{n-s-k}}{s! \, k! \, (n-s-k)!}$$

$$= n(n-1)p_1 p_2 (p_1 + p_2 + p_3)^m = n(n-1)p_1 p_2$$

(where $m = n - 2$, $s = i - 1$, $k = j - 1$). Hence $\operatorname{Cov}(X, Y) = -np_1 p_2$, and in this case $\operatorname{Cov}(X, Y) = -1/3$.

182. We have

$$\frac{P[X_1 = n_1, \ldots, X_i = n_i, \ldots, X_j = n_j, \ldots, X_k = n_k]}{P[X_1 = n_1, \ldots, X_i = n_i + 1, \ldots, X_j = n_j - 1, \ldots, X_k = n_k]} = \frac{n_i + 1}{n_j} \cdot \frac{p_j}{p_i}$$

$$\text{for} \quad i, j = 1, 2, \ldots, k, \quad i = j.$$

Thus the vector $(n_1^0, n_2^0, \ldots, n_{k+1}^0)$ is the most probable, if and only if, for every pair of indices i, j ($i \ne j$) the inequality

$$p_i n_j^0 \le p_j (n_i^0 + 1) \tag{1}$$

is satisfied. From this, summing over j with $j \ne i$ we have

$$p_i \sum_{\substack{j=1 \\ j \ne i}}^{k+1} n_j^0 \le (n_i^0 + 1) \sum_{\substack{j=1 \\ j \ne i}}^{k+1} p_j,$$

$$p_i(n - n_i^0) \le (n_i^0 + 1)(1 - p_i), \tag{2}$$

$$np_i \le n_i^0 - p_i + 1 < n_i^0,$$

$$np_i - 1 \le n_i^0 - p_i < n_i^0.$$

Similarly, summing over i with $j \ne i$ we get

$$n_j^0 \le p_j(n + k), \qquad j = 1, 2, \ldots, k + 1. \tag{3}$$

From (2) and (3) the result follows.

183. Putting $np_i = \lambda_i$ ($i = 1, 2, \ldots, k$), in the multinomial probability function, we have

$$P[X_i = n_i, i = 1, 2, \ldots, k]$$

$$= \frac{\lambda_1^{n_1} \lambda_2^{n_2} \ldots \lambda_k^{n_k}}{n_1! \, n_2! \ldots n_k!} \cdot \frac{n(n-1) \ldots (n+1-n_0)}{n^{n_1} n^{n_2} \ldots n^{n_k}} \cdot \left(1 - \frac{\sum_{j=1}^{k} \lambda_j}{n}\right)^{n-n_0}$$

$$= \frac{\lambda_1^{n_1} \lambda_2^{n_2} \dots \lambda_k^{n_k}}{n_1! \, n_2! \dots n_k!} \cdot 1 \cdot \left(1 - \frac{1}{n}\right)\left(1 - \frac{n_0 - 1}{n}\right) \frac{\left(1 - \frac{\sum \lambda_j}{n}\right)^n}{\left(1 - \frac{\sum \lambda_j}{n}\right)^{n_0}}. \quad (1)$$

where we set $n_0 = \sum_{j=1}^{k} n_j$.

But for $n \to \infty$ we have

$$\left(1 - \frac{S}{n}\right) \to 1, \qquad S = 1, 2, \dots, n_0 - 1,$$

$$\left(1 - \frac{\sum_{j=1}^{k} \lambda_j}{n}\right)^{n_0} \to 1,$$

$$\left(1 - \frac{\sum_{j=1}^{k} \lambda_j}{n}\right)^{n} \to \exp\left(-\sum_{j=1}^{k} \lambda_j\right).$$

Hence from (1) the result follows by taking the limit $(n \to \infty)$.

184. (a)

$$F(x, y) = 2 \int_0^x du \int_0^y (1 + u + v)^{-3} \, dv = \int_0^x [(1 + u)^{-2} - (1 + u + y)^{-2}] \, du$$

$$= 1 - \frac{1}{1 + x} + \frac{1}{1 + x + y} - \frac{1}{1 + y}.$$

(b)

$$f_X(x) = \int_0^\infty f(x, y) \, dy = 2 \int_0^\infty (1 + x + y)^{-3} \, dy = \frac{1}{(1 + x)^2}.$$

(c)

$$f_Y(y \mid X = x) = \frac{f(x, y)}{f_X(x)} = \frac{2(1 + x)^2}{(1 + x + y)^3}.$$

185. $f_X(x) \, dx \sim$ area of the infinitesimal strip $ABB'A'$ (see figure), so that

$$f(x) \, dx = \frac{1}{\pi}(2\sqrt{1 - x^2}).$$

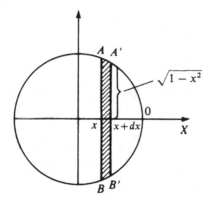

Hence

$$f_X(x) = \frac{2}{\pi}\sqrt{1 - x^2}, \qquad f_Y(y) = f_X(y), \qquad -1 < y < 1,$$

$$F_X(x) = \frac{1}{2} + \frac{1}{\pi}(x\sqrt{1 - x^2} + \arcsin x), \qquad 0 \le x \le 1,$$

$$= 1 - \frac{1}{\pi}(\text{area of the circular part } (AOB)),$$

$$F_Y(y) = F_X(y) = 1 - F_X(-y).$$

X and Y are not independent because, for example,

$$P\left[X > \frac{\sqrt{2}}{2}, Y > \frac{\sqrt{2}}{2}\right] = 0 \ne P\left[X > \frac{\sqrt{2}}{2}\right]P\left[X > \frac{\sqrt{2}}{2}\right],$$

that is, the relation

$$P[X \in A, Y \in B] = P[X \in A]P[Y \in B]$$

does not hold for all A, B.

186. $$P[X < Y < Z] = \int_0^1 dz \int_0^z dy \int_0^y dx = \frac{1}{6}.$$

This is so, since every ordering of X, Y, Z has the same probability because of symmetry of the joint density.

187. (a)

$$F_X(x) = F_Y(x) = x^2, \quad F(x, y) = x^2y^2, \quad f_X(x) = f_Y(y) = 2x, \quad 0 < x, \quad y < 1. \tag{1}$$

$$f_X(x \mid Y = y) = \frac{4xy}{2y} = 2x, \qquad f_Y(y \mid X = x) = 2y. \tag{2}$$

From (1), the relations (2) follow because of the independence of X, Y.

(b)
$$f_X(x) = \frac{e^{-x}}{8}\int_{-x}^x (x^2 - y^2)\, dy = \frac{x^3 e^{-x}}{6}, \qquad x > 0,$$

$$f_Y(y) = \frac{1}{8}\int_{|y|}^\infty (x^2 - y^2)e^{-x}\, dx = \tfrac{1}{4}e^{-|y|}(1 + |y|),$$

$$F_X(x) = \frac{1}{6}\int_0^x u^3 e^{-u}\, du = 1 - \left(e^{-x} + xe^{-x} + \frac{x^2}{2} + \frac{x^3}{3!}\right),$$

$$F_Y(y) = 1 - \tfrac{1}{2}e^{-y}\left(1 + \frac{y}{2}\right), \qquad y > 0,$$

$$F_Y(y) = 1 - F_Y(-y), \qquad y < 0.$$

188. We have

$$\mu_x = E(X) = -0.35 + 0.20 = -0.15,$$

$$E(X^2) = 0.35 + 0.20 = 0.55,$$

$$\text{Var}(X) = E(X^2) - \mu_x^2 = 0.55 - 0.0225 = 0.5275,$$

$$\sigma_x = \sqrt{\text{Var}(X)} = 0.726,$$

$$\mu_y = E(Y) = -0.3 + 0.1 + 0.8 = 0.6,$$

$$E(Y^2) = 0.3 + 0.1 + 1.6 = 2, \qquad \text{Var}(Y) = E(Y^2) - \frac{2}{y} = 2 - 0.36 = 1.64,$$

$$\sigma_y = \sqrt{\text{Var}(Y)} = 1.28,$$

$$E(XY) = 0.1 - 0.05 - 0.2 + 0.2 = 0.05,$$

$$\rho = \frac{\text{Cov}(X, Y)}{\sigma_x \sigma_y} = \frac{E(XY) - \mu_x \mu_y}{\sigma_x \sigma_y} = \frac{0.05 + 0.09}{0.726 + 1.286} = \frac{0.14}{0.932} = 0.15.$$

(a) The regression lines of X on Y and of Y on X are given by

$$X - \mu_x = \rho \frac{\sigma_x}{\sigma_y}(y - \mu_y), \tag{1}$$

$$Y - \mu_y = \rho \frac{\sigma_y}{\sigma_x}(x - \mu_x), \tag{2}$$

respectively. Substituting parameters in (1) and (2), we have

$$x + 0.15 = 0.084(y - 0.6), \tag{3}$$

$$y - 0.6 = 0.265(x + 0.15). \tag{4}$$

Observe that both lines pass through the point ($\mu_x = -0.15$, $\mu_y = 0.6$); moreover, line (3) passes through ($x = -0.2$, $y = 0$) and line (4) passes through ($x = 0$, $y = 0.64$) (see figure).

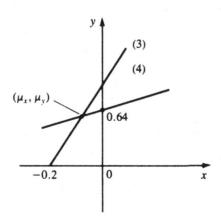

(b)

$$E(Y|X = x) = m(x), \qquad m(-1) = 0.29, \quad m(0) = 0.78, \quad m(1) = 0.75.$$

189. If X and Y are independent, then

$$p_{ij} = P[X = x_i]P[Y = y_j] = p_i q_j, \qquad i, j = 1, 2, \ldots,$$

and the nth line is $(p_n q_1, p_n q_2, \ldots) = p_n \mathbf{q}$ $(n = 1, 2, \ldots)$. So there is only one linearly independent row. Hence the rank $r(p)$ of the matrix P is 1. Conversely, if $r(p) = 1$, then there is a vector $\mathbf{a} = (a_1, a_2, \ldots)$ so that every row γ_n can be written as

$$\gamma_n = \lambda_n \mathbf{a}, \qquad n = 1, 2, \ldots \quad (\lambda_n = \text{constant}).$$

But

$$\sum_i p_{ij} = q_j \qquad \text{gives} \qquad \sum_n \gamma_n = (q_1, q_2, \ldots) = \mathbf{q} = \mathbf{a}\left(\sum_n \lambda_n\right). \qquad (1)$$

So $\mathbf{a}$ is a multiple of $\mathbf{q}$ and there are constants λ_n^* so that

$$p_{nj} = \lambda_n^* q_j, \qquad n, j = 1, 2, \ldots.$$

Hence X and Y are independent (λ_n^* will be equal to p_n necessarily).

190. Consider the random variables X and Y with joint probability function

$$P[X = x_i, Y = y_j] = p_{ij}, \qquad i, j = 1, 2, \qquad (1)$$

and marginal probability functions

$$P[X = x_i] = p_i, \quad i = 1, 2, \qquad \text{and} \qquad P[Y = y_j] = q_j, \quad j = 1, 2, \qquad (2)$$

respectively. We have

$$\sum_{i=1}^{2} \sum_{j=1}^{2} p_{ij} = 1, \qquad (3a)$$

$$\sum_{j=1}^{2} p_{ij} = p_i, \qquad i = 1, 2, \qquad (3b)$$

$$\sum_{i=1}^{2} p_{ij} = q_j, \qquad j = 1, 2, \qquad (3c)$$

$$p_1 + p_2 = 1, \qquad q_1 + q_2 = 1. \qquad (4)$$

Since X and Y are supposed to be orthogonal we get

$$\text{Cov}(X, Y) = E(XY) - E(X)E(Y) = 0, \qquad (5)$$

$$E(XY) = \sum_{i=1}^{2} \sum_{j=1}^{2} x_i y_j p_{ij}, \qquad E(X) = \sum_{i=1}^{2} x_i p_i, \qquad E(Y) = \sum_{j=1}^{2} y_j q_j.$$

Hence (5) is written

$$x_1 y_1 (p_{11} - p_1 q_1) + x_1 y_2 (p_{12} - p_1 q_2) + x_2 y_1 (p_{21} - p_2 q_1) + x_2 y_2 (p_{22} - p_2 q_2)$$
$$= 0. \qquad (6)$$

From (3) we have

$$p_{12} = p_1 - p_{11}, \qquad p_{21} = q_1 - p_{11}, \qquad p_{22} = q_2 - p_{12} = p_{11}p_1 + p_2. \quad (7)$$

Replacing them in (6) we get

$$x_1 y_1 (p_{11} - p_1 q_1) - x_1 y_1 (p_{11} - p_1 + p_1 q_2) - x_2 y_1 (p_{11} - q_1 + p_2 q_1)$$
$$+ x_2 y_2 (p_{11} - p_1 - q_2 - p_2 q_2) = 0.$$

The above equation, because of (4), is written

$$(x_1 y_1 - x_1 y_2 - x_2 y_1 + x_2 y_2)(p_{11} - p_1 q_1) = 0,$$

or

$$(x_1 - x_2)(y_1 - y_2)(p_{11} - p_1 q_1) = 0,$$

and since $x_1 \neq x_2$, and $y_1 \neq y_2$, it follows that $p_{11} = p_1 q_1$. Equations (7), because of the last relation, are written

$$p_{12} = p_1 - p_1 q_1 = p(1 - q_1) = p_1 q_2,$$
$$p_{21} = q_1 - p_1 q_1 = q_1(1 - p_1) = p_2 q_1,$$
$$p_{22} = p_2 - p_{21} = p_2 - p_2 q_1 = p_2(1 - q_1) = p_2 q_2,$$

that is, $p_{ij} = p_i q_j$ $(i, j = 1, 2)$. So X, Y are independent.

191. (a)

$$P(t) = (pt + q)^n, \qquad M(t) = P(e^t) = (pe^t + q)^n, \qquad \varphi(t) = M(it) = (pe^{it} + q)^n,$$
$$\mu = p'(1) = M'(0) = i^{-1}\varphi'(0) = np, \qquad \sigma^2 = npq.$$

(b)

$$P(t) = e^{\lambda(t-1)}, \qquad M(t) = e^{\lambda(e^t-1)}, \qquad \varphi(t) = e^{\lambda(e^{it}-1)}, \qquad \mu = \sigma^2 = \lambda.$$

(c)

$$P(t) = \frac{pt}{1 - qt}, \qquad M(t) = \frac{pe^t}{1 - qe^t}, \qquad \varphi(t) = \frac{pe^{it}}{1 - qe^{it}}, \qquad \mu = \frac{1}{p}, \qquad \sigma^2 = \frac{q}{p^2}.$$

(d)

$$P(t) = \left(\frac{p}{1 - qt}\right)^r, \quad M(t) = \left(\frac{p}{1 - qe^t}\right)^r, \quad \varphi(t) = \left(\frac{p}{1 - qe^{it}}\right)^r, \quad \mu = \frac{1}{p}, \quad \sigma^2 = \frac{rq}{p^2}.$$

192. The characteristic function of X is

$$\varphi(t) = \frac{1}{\pi} \int_{-\infty}^{\infty} \frac{e^{itx}}{1 + x^2} \, dx = e^{-|t|}.$$

Hence

$$\varphi_{X_1 + X_2}(t) = \varphi_{2X}(t) = \varphi_X(2t) = e^{-|2t|} = \varphi_{X_1}(t)\varphi_{X_2}(t),$$

that is, (6.11) is only a necessary but not sufficient condition for the independence of two random variables.

193. $P_Y(t) = E(t^Y) = E(t^{3X+2}) = t^2 E(t^{3X}) = t^2 P_X(t^3).$

194. $\text{Var}(S_N) = P''_{S_N}(1) + P'_{S_N}(1) - P'_{S_N}(1)^2,$ (1)

where

$$P_{S_N}(t) = P_N(P_X(t)) \quad \text{(see (6.4))},$$
$$P'_{S_N}(t) = P'_X(t)P'_N(P_X(t)),$$
$$P''_{S_N}(t) = P''_X(t)P'_N(P_X(t)) + P'_X(t)^2 P''_N(P_X(t)).$$

Hence we have

$$P'_{S_N}(t) = E[S_N] = P'_X(1)P'_N(P_X(1)) = P'_X(1)P'_N(1) = E(X)E(N),$$
$$P''_{S_N}(1) = P''_1(1)P'_N(1) + [P'(1)]^2 P''_N(1)$$
$$= E[X(X-1)]E(N) + E(N^2)E[X(X-1)].$$

Replacing in (1) we have the result.

195. The number of the blond children is

$$S_N = \sum_{i=1}^{N} X_i.$$

The number of children N is geometric with parameter $p_1 = 1/2$. The X_i's are independent Bernoulli with parameter p. Because of (6.4) and Exercise 191 we have as the probability genrating function of S_N

$$P_{S_N}(t) = \frac{p_1(pt + q)}{1 - q_1(pt + q)} = \frac{at + b}{1 - ct},$$

where $a = p_1 p_2/(1 - qq_1)$, $b = qq_1/(1 - qq_1)$, $c = q_1 p/(1 - qq_1)$. The probability P_k of k blond children is given by the coefficient of t^k in the expansion of $P_{S_N}(t)$. This gives

$$P_0 = b, \qquad P_k = c^{k-1}(a + bc), \qquad k = 1, 2, \dots.$$

This is a geometric distribution with modified first term.

196. Let

$$X = (X_1, X_2, X_3),$$

where

$$X_{jk} = \sum_{j=1}^{N} X_{jk}$$

and, for every $k = 1, 2, \dots, N$, $P[X_{jk} = 1] = p_j = 1 - P[X_{jk} = 0]$ $(j = 1, 2, 3)$. The probability generating function of $\xi_k = (X_{1k}, X_{2k}, X_{3k})$ for $k = 1, 2, \dots,$

N, is

$$P_k(t_1, t_2, t_3) = p_1 t_1 + p_2 t_2 + p_3 t_3 + p_4, \qquad k = 1, 2, \ldots, N.$$

The random variable N is, by assumption, Poisson with parameter $\lambda = 150$. It can be shown that the probability generating function of $X = (X_1, X_2, X_3) = \sum_{k=1}^N \xi_k$. Similarly the probability generating function of S_N (cf. (6.4)) is given by

$$P_X(t_1, t_2, t_3) = P_N(P_k(t_1, t_2, t_3)) = e^{\lambda(p_k - 1)}$$

$$= e^{\lambda(p_1 t_1 + p_2 t_2 + p_3 t_3 + p_4 - 1)} = \prod_{i=1}^3 e^{\lambda p_i(t_i - 1)}.$$

This is a product of Poisson probability generating functions with parameters λp_i. Then X_1, X_2, X_3 are independent Poisson random variables with parameters $\lambda p_1, \lambda p_2, \lambda p_3$, respectively. This can be directly shown as follows:

$$P[X_1 = k_1, X_2 = k_2, X_3 = k_3] = e^{-\lambda} \sum_{n=k_1+k_2+k_3}^{\infty} \frac{\lambda^n}{n!} \frac{n!}{k_1! \, k_2! \, k_3! \, k_4!} p_1^{k_1} p_2^{k_2} p_3^{k_3} p_4^{k_4}$$

$$= e^{-\lambda} \frac{(\lambda p_1)^{k_1} (\lambda p_2)^{k_2} (\lambda p_3)^{k_3}}{k_1! \, k_2! \, k_3!} \sum_{k_4=0}^{\infty} \frac{(\lambda p_4)^{k_4}}{k_4!}$$

$$= \prod_{i=1}^3 e^{-\lambda p_i} \frac{(\lambda p_i)^{k_i}}{k_i!}.$$

197. We have

$$f_X(u) = \tfrac{1}{2}[f_{X_1}(u) + f_{X_2}(u)],$$

where

$$f_{X_1}(u) = \frac{a - a \cos(u/a)}{\pi u^2}, \qquad f_{X_2}(u) = \frac{b - b \cos(u/b)}{\pi u^2}.$$

We have

$$\varphi_Y(t) = \frac{c}{\pi} \int_{-\infty}^{\infty} e^{itu} \frac{1 - \cos(u/c)}{u^2} \, du = \begin{cases} 1 - c|t|, & |t| \leq 1/c, \\ 0 & \text{for } |t| > 1/c, \end{cases}$$

and

$$\varphi_{X_1}(t) = \begin{cases} 1 - a|t|, & |t| \leq 1/a, \\ 0, & |t| > 1/a, \end{cases} \qquad \varphi_{X_2}(t) = \begin{cases} 1 - b|t|, & |t| \leq 1/b, \\ 0, & |t| > 1/b; \end{cases}$$

hence, for $2c = a + b$

$$\varphi_X(t) = \varphi_Y(t) \qquad \text{for} \quad |t| \leq \min\{a^{-1}, b^{-1}, c^{-1}\} = \min\{a^{-1}, b^{-1}\}.$$

Note. The density

$$f(u) = \frac{1 - \cos(\lambda u)}{\lambda \pi u^2}, \qquad -\infty < u < \infty,$$

has characteristic function

$$\varphi(t) = \begin{cases} 1 - |t|/\lambda, & |t| \le \lambda, \\ 0, & |t| > \lambda. \end{cases}$$

198.
$$E(S_n) = \sum_{i=1}^{n} E(X_k^2) = n(1 + \mu^2),$$

$$\tag{1}$$

$$\text{Var}(S_n) = \sum_{k=1}^{n} \text{Var}(X_k^2) = n[E(X_k^4) - E(X_k^2)^2].$$

But if $Z \sim N(0, 1)$, then $X_k = Z + \mu$ and

$$E(X_k^4) = E(Z + \mu)^4 = E(Z^4) + 4\mu E(Z^3) + 6\mu^2 E(Z^2) + 6\mu^3 E(Z) + \mu^4$$
$$= 3 + 6\mu^2 + \mu^4, \tag{2}$$

since

$$E(Z^k) = i^{-k}\varphi_Z^{(k)}(0) = \begin{cases} 0 & \text{for } k = 2m + 1, \\ 1, 3, \dots, (2m - 1) & \text{for } k = 2m \ (m = 1, 2, \dots), \end{cases}$$

where

$$\varphi_Z(t) = e^{-t^2/2}.$$

From (1) and (2) we have

$$\text{Var}(S_n) = n[(3 + 6\mu^2 + \mu^4) - (1 + \mu^2)^2] = 2n(1 + 2\mu^2).$$

On the other hand, we have

$$E(\chi_v^2) = v, \qquad \text{Var}(\chi_v^2) = 2v, \qquad E(T) = av, \qquad \text{Var}(T) = 2a^2 v.$$

Setting

$$E(T) = E(S_n), \qquad \text{Var}(T) = \text{Var}(S_n)$$

we have

$$a = \frac{1 + 2\mu^2}{1 + \mu^2}, \qquad v = \frac{n(1 + \mu^2)^2}{1 + 2\mu^2}.$$

199. By an extension of the theorem of total probability for the expected values, we have

$$\varphi_{S_N}(t) = E(e^{itS_N}) = \sum_{n=0}^{\infty} E[e^{itS_N}|N = n]P[N = n]$$
$$= \sum_n E(e^{itS_n})P[N = n] = \sum_n \varphi_{X_i}^n(t)P[N = n] = P_N(\varphi_{X_i}(t)),$$

because for given $N = n$ the characteristic function of $S_n = X_1 + \cdots + X_n$ is $\varphi_{X_i}^n(t)$.

200. The probability generating function of a Poisson distribution is

$$P(s) = e^{\lambda(s-1)}, \tag{1}$$

the characteristic function of the Cauchy distribution with density

$$f(y) = \frac{1}{\pi} \frac{1}{1 + y^2}$$

is

$$\varphi(t) = e^{-|t|}. \tag{2}$$

So the characteristic function distribution of S_N,

$$S_N = X_1 + X_2 + \cdots + X_N,$$

where $X_j (j = 1, 2, \ldots, N)$, are independent and identically distributed Cauchy random variables and N is Poisson independent from X_j, is

$$\varphi_{S_N}(t) = P(\varphi(t)) = e^{\lambda(e^{-|t|} - 1)}.$$

The density of S_N is

$$f(x) = \frac{1}{2\pi} \int_{-\infty}^{+\infty} e^{-itx} e^{\lambda(e^{-|t|} - 1)} \, dt = \frac{e^{-\lambda}}{\pi} \int_0^\infty e^{\lambda e^{-t}} \cos tx \, dx$$

$$= \frac{e^{-\lambda}}{\pi} \int_0^\infty \left(\sum_{n=0}^\infty \frac{(\lambda e^{-t})^n}{n!} \right) \cos tx \, dx = \frac{e^{-\lambda}}{\pi} \sum_{n=0}^\infty \frac{\lambda^n}{n!} \int_0^\infty \cos tx e^{-nt} \, dt$$

$$= \frac{e^{-\lambda}}{\pi} \int_0^\infty \cos tx \, dx + \frac{1}{\pi} e^{-\lambda} \sum_{n=1}^\infty \frac{\lambda^n}{n!} \frac{n}{n^2 + x^2}$$

$$= \frac{1}{\pi} e^{-\lambda} \int_0^\infty \cos xt \, dt + \frac{1}{\pi} e^{-\lambda} \sum_{n=1}^\infty \frac{\lambda^n}{n!} \frac{n}{n^2 + x^2}.$$

We observe that this is a mixture of Cauchy distributions with weights W_n, i.e., of the form

$$\sum_n W_n f_n(x),$$

where

$$W_n = e^{-\lambda} \frac{\lambda^n}{n!}, \qquad n = 1, 2, \ldots.$$

The $f_n(x)$ represent Cauchy densities

$$f_n(x) = \frac{1}{\pi} \frac{n}{n^2 + x^2}, \qquad -\infty < x < \infty, \quad n = 1, 2, \ldots,$$

$$f_0(x) = \frac{1}{\pi} \int_0^\infty \cos tx \, dt.$$

201. We have (see Exercise 191), putting $p = \lambda/n$,

$$\varphi_{Z_n}(t) = (pe^{it} + q)^n = \left[1 + \frac{\lambda(e^{it} - 1)}{n} \right]^n_{n + \infty} \to e^{\lambda(e^{it} - 1)},$$

that is, the characteristic function of Poisson with parameter λ. According to the continuity theorem for characteristic functions, the $F_{Z_n}(x) \to F_Z(x)$. Hence the approximation of a binomial by Poisson, for $p \to 0, n \to \infty$, so that $np \to \lambda$, is

$$\binom{n}{k} p^k q^{n-k} \approx e^{-\lambda} \frac{\lambda^k}{k!}, \qquad k = 0, 1, 2, \dots .$$

202. We have

$$\varphi(t) = \frac{\lambda^s}{\Gamma(s)} \int_0^\infty e^{itx} e^{-\lambda x} x^{s-1} \, dx = \frac{\lambda}{\Gamma(s)} \int_0^\infty x^{s-1} e^{-(\lambda - it)x} \, dx$$

$$= \frac{\lambda^s}{(\lambda - it)^s} = \frac{1}{\left(1 - \dfrac{it}{\lambda}\right)^s}.$$

For $\lambda = 1/2$, $s = n/2$, the Gamma distribution gives χ_n^2. Hence

$$\varphi_{\chi_n^2}(t) = (1 - 2it)^{-n/2}.$$

203. Show that the characteristic function $\varphi_{Y_n}(t)$ of Y_n,

$$\varphi_{Y_n}(t) = \prod_{k=1}^n \cos(t/2^k) = \prod_{k=1}^n \frac{\sin(t/k)}{2 \sin(t/2^{k+1})},$$

tends (as $n \to \infty$) to $\sin t/t$.

204.
$$\varphi_X(t) = \int_{-\infty}^\infty e^{itx} [\lambda \, dF_1(x) + (1 - \lambda) \, dF_2(x)]$$

$$= \lambda \varphi_1(t) + (1 - \lambda)\varphi_2(t),$$

where φ_j is the characteristic function of F_j ($j = 1, 2$). For $F_j \sim N(\mu_j, \sigma_j^2)$ we have

$$\varphi_j(t) = \exp(i\mu_j t - \tfrac{1}{2}\sigma_j^2 t^2), \qquad j = 1, 2,$$

hence

$$\varphi_X(t) = \lambda \exp(i\mu_1 t - \tfrac{1}{2}\sigma_1^2 t^2) + (1 - \lambda) \exp(i\mu_2 t - \tfrac{1}{2}\sigma_2^2 t^2),$$

$$E(X) = i^{-1}\varphi_X'(0) = \lambda \mu_1 + (1 - \lambda)\mu_2,$$

$$\text{Var}(X) = -\varphi_X''(0) + [\varphi_X'(0)]^2 = \lambda \sigma_1^2 + (1 - \lambda)\sigma_2^2.$$

205. The exponetial is a Gamma distribution (see Exercise 202) with parameters $s = 2$ and $\lambda = \vartheta$. Hence

$$\varphi(t) = \left(1 - \frac{it}{\vartheta}\right)^{-1}$$

and the generating function of cumulants $\Psi(t)$ is given by

$$\Psi(t) = \log \varphi(t) = -\log\left(1 - \frac{it}{\vartheta}\right) = \frac{it}{\vartheta} + \frac{(it)^2}{2\vartheta^2} + \cdots + \frac{(it)^r}{r\vartheta^r} + \cdots .$$

Then the cumulant of order r is

$$\kappa_r = \frac{(r-1)!}{\vartheta^r}.$$

206. By the definition of Y we have

$$P[Y = (2k+1)h/2] = e^{-kh\vartheta}[1 - e^{-\vartheta h}]$$

and hence the probability generating function of Y is

$$P_Y(t) = E(t^Y) = \sum_{k=0}^{\infty} t^{(2k+1)h/2} e^{-kh\vartheta}[1 - e^{-\vartheta h}]$$

$$= t^{h/2}(1 - e^{-\vartheta h})/(1 - t^h e^{-\vartheta h}).$$

From the above relation we have

$$P_Y'(1) = E(Y) = \frac{1}{\vartheta} + \left[\frac{1}{2}h \coth\left(\frac{1}{2}h\vartheta\right) - \frac{1}{\vartheta}\right] > \frac{1}{\vartheta} = E(X)$$

because of $0 < \tanh(\frac{1}{2}\vartheta h) < \frac{1}{2}\vartheta h$ and

$$\mathrm{Var}(Y) = \frac{1}{\vartheta^2} - \frac{1}{\vartheta^2}\left[1 - \left(\frac{1}{4}\vartheta^2 h^2 \Big/ \sinh^2\left(\frac{1}{2}\vartheta h\right)\right)\right] < \frac{1}{\vartheta^2} = \mathrm{Var}(X)$$

because of $\sin(h\vartheta/2) > \frac{1}{2}\vartheta h$.

207. We observe that by the orthogonal transformation

$$t^* = t \cos\vartheta + u \sin\vartheta, \qquad u^* = t \sin\vartheta - u \cos\vartheta,$$

$$x^* = x \cos\vartheta + y \sin\vartheta, \qquad y^* = x \sin\vartheta - y \cos\vartheta, \qquad (1)$$

with Jacobian -1 we have

$$\varphi(t, u) = \int_{-\infty}^{\infty}\int_{-\infty}^{\infty} e^{i(tx+uy)} g(x^2 + y^2)\, dx\, dy$$

$$= \int_{-\infty}^{\infty}\int_{-\infty}^{\infty} e^{i(t^*x^*+u^*y^*)} g(x^{*2} + y^{*2})\, dx^*\, dy^*.$$

That is, $\varphi(t, u)$ is invariant under the orthogonal transformation (1) and so it must be a <u>function</u> of the invariant (function) of the transformation, i.e., of the length $\sqrt{t^2 + u^2}$ of (t, u). Then we deduce

$$\varphi(t, u) = \varphi^*(t^2 + u^2). \qquad (2)$$

For independent random variable(s) X and Y with density $g(x^2 + y^2)$ we have

$$\varphi_{X,Y}(t, u) = \varphi_1(t)\varphi_2(u).$$

Because of (2) we have

$$\varphi^*(t^2 + u^2) = \varphi_1(t)\varphi_2(u), \qquad (3)$$

where φ_1, φ_2 are the characteristic function(s) of X, Y, respectively. We can write relation (3) in the form

$$\varphi(r) = \varphi_1(t)\varphi_2(u), \qquad r^2 = u^2 + v^2, \tag{4}$$

then putting $t = 0$ we deduce that $\varphi = \varphi_2$ and putting $u = 0$ we deduce that $\varphi = \varphi_1$. Hence

$$\varphi_1(t) = \varphi_1^*(t^2), \qquad \varphi_2(t) = \varphi_2^*(u^2).$$

So φ satisfies the functional equation

$$\varphi(t^2 + v^2) = \varphi(t^2)\varphi(u^2)$$

the only solution of which (see, e.g., Feller, 1957) is $\varphi(x) = e^{\beta x}$. Since φ is a characteristic function, $|\varphi(t)| \leq 1$, and hence $\beta < 0$. So for some $\sigma^2 > 0$ we have

$$\varphi(t^2) = e^{-\sigma^2 t^2/2},$$

that is the characteristic function of the normal $N(0, \sigma^2)$.

208. Let

$$U = \frac{1}{n}\sum_{j=1}^{n}(X_j - Y_j), \qquad V = \frac{1}{n}\sum_{j=1}^{n}(X_j + Y_j).$$

The characteristic functions of U and V are

$$\varphi_U(t) = \frac{1}{\left(1 + \dfrac{t^2}{n}\right)^{2n}}, \qquad \varphi_V(t) = \frac{1}{\left(1 + \dfrac{t^2}{n}\right)^{2n}}.$$

Hence U, V are identically distributed, having the same characteristic function for every t. The joint characteristic function of U and V is

$$\varphi(t_1, t_2) = E\exp[i(t_1 U + t_2 V)] = E\exp\left[i\left(\frac{t_1 + t_2}{n}\sum_{j=1}^{n}X_j + \frac{t_1 - t_2}{n}\sum_{j=1}^{n}Y_j\right)\right]$$

$$= \left[1 + \left(\frac{t_1 + t_2}{n}\right)^2\right]^{-n}\left[1 + \left(\frac{t_1 - t_2}{n}\right)^2\right]^{-n}.$$

We observe that

$$\varphi(t_1, t_2) \neq \varphi(t_1, 0)\varphi(0, t_2) = \varphi_U(t_1)\varphi_V(t_2).$$

So U, V are not independent. It is easily shown that $\text{Cov}(U, V) = 0$, that is, U and V are uncorrelated. This can also be shown from the fact that the coefficient of $t_1 t_2$ in the expansion of $\log \varphi(t_1, t_2)$ is 0.

209. (a) Let X_1, X_2 be independent and identically distributed random variables with characteristic function $\varphi(t)$. Then the characteristic function of $X_1 - X_2$ is $\varphi(t)\overline{\varphi(t)} = |\varphi(t)|^2$.

(b) Let $X_1, X_2, \ldots, X_n$ be independent and identically distributed random variables with characteristic function $\varphi(t)$ and let N be Poisson with parameter λ, independent of the X_j's. Then, from Exercise 199, the characteristic function of the sum

$$S_N = X_1 + X_2 + \cdots + X_N$$

is $e^{\lambda(\varphi(t)-1)}$.

210. Similarly, as in Exercise 196, the multinomial random variable X is written

$$X = (X_1, X_2, \ldots, X_k) = \sum_{j=1}^{k} \xi_j,$$

where the probability generating function of ξ_j is

$$P_j(t_1 t_2, \ldots, t_k) = p_1 t_1 + p_2 t_2 + \cdots + p_k t_k + p_{k+1}.$$

Hence the probability generating function of X is

$$P(t_1, t_2, \ldots, t_k) = (p_1 t_1 + p_2 t_2 + \cdots + p_k t_k + p_{k+1})^n.$$

From the above relation we have

$$E(X_j) = \frac{\partial P}{\partial t_1}\bigg|_{t_1=t_2=\cdots=t_k=1} = np_j, \quad j = 1, 2, \ldots, k,$$

$$E(X_j X_s) = \frac{\partial^2 P}{\partial t_j\, \partial t_s}\bigg|_{t_1=t_2=\cdots=t_k=1},$$

$$E[X_j(X_j - 1)] = \frac{\partial^2 P}{\partial t_j^2}\bigg|_{t_1=t_2=\cdots=t_k=1}, \quad j = 1, 2, \ldots, k,$$

hence the assertion is proved.

211. Putting λ_j/n instead of p_j into $P(t_1, \ldots, t_k)$ of Exercise 210 we have

$$P(t_1, \ldots, t_k) = \left(1 + \frac{\sum_{j=1}^{k} \lambda_j(t_j - 1)}{n}\right)^n \xrightarrow[n\to\infty]{} \exp\left[\sum_{j=1}^{n} \lambda_j(t_j - 1)\right] = \prod_{j=1}^{k} e^{\lambda_j(t_j-1)},$$

and hence the assertion is proved.

212. We have

$$\varphi(t_1, \ldots, t_k) = \sum_{n=0}^{\infty} \sum_{n_1+\cdots+n_k=n} \left(1 + \sum_{j=1}^{k} \vartheta_j\right) \frac{\Gamma(s + n)}{n!\,\Gamma(s)} \left(\sum_{j=1}^{k} \vartheta_j e^{it_j}\right)^n$$

$$= \left[1 + \sum_{j=1}^{\vartheta} \vartheta_j(1 - e^{it_j})\right]^{-s}.$$

By virtue of (2.13), we find

$$E(X_j) = s\vartheta_j, \quad \text{Var}(X_j) = s\vartheta_j(1 + \vartheta_j), \quad \text{Cov}(X_j, X_k) = s\vartheta_j\vartheta_k.$$

213. The conditional characteristic function of X, given $Y = y$, is

$$\varphi_X(t_1 | Y = y) = E[e^{it_1 X} | Y = y] = \int_{-\infty}^{\infty} e^{it_1 X} f_X(x | Y = y)\, dx,$$

while the joint characteristic function of X and Y is

$$\varphi(t_1, t_2) = E[e^{i(t_1 X + t_2 Y)}] = E[E(e^{i(t_1 X + t_2 Y)} | Y)]$$

$$= E[e^{it_2 Y} E(e^{it_1 X} | Y)] = E[e^{it_2 Y} \varphi_X(t_1 | Y)]$$

$$= \int_{-\infty}^{\infty} e^{it_2 y} \varphi_X(t_1 | Y = y) f_Y(y)\, dy.$$

By the inversion formula we have

$$\varphi_X(t_1 | Y = y) f_Y(y) = \frac{1}{2\pi} \int_{-\infty}^{\infty} e^{-iyt_2} \varphi(t_1, t_2)\, dt_2 \tag{1}$$

and since

$$f_Y(y) = \frac{1}{2\pi} \int_{-\infty}^{\infty} e^{it_2 y} \varphi(0, t_2)\, dt_2,$$

the result follows.

214. The joint characteristic function of Y, Z is

$$\varphi^*(t_1, t_2) = E[\exp(i(t_1 Y + t_2 Z))] = E[\exp(it_1(S_1 + S_3) + it_2(S_2 + S_3))]$$

$$= E \exp(i[t_1 S_1 + (t_1 + t_2)S_3 + t_2 S_2])$$

$$= [\varphi(t_1)]^{n_1 - n} [\varphi(t_1 + t_2)]^n [\varphi(t_2)]^{n_2 - n},$$

where $\varphi(t)$ is the characteristic function of X. By virtue of (1) of Exercise 213, we have

$$E(Y | Z = z) f_Z(z) = \frac{1}{2\pi} \int_{-\infty}^{\infty} e^{-it_2 z} \left[i^{-1} \frac{\partial \varphi^*}{\partial t_1} \right]_{t_1 = 0} dt_2, \tag{1}$$

$$E(Y^2 | Z = z) f_Z(z) = \frac{1}{2\pi} \int_{-\infty}^{\infty} e^{-it_2 z} \left[i^{-2} \frac{\partial \varphi^*}{\partial t_1} \right]_{t_1 = 0} dt_2, \tag{2}$$

given that the density of Z is

$$f_Z(z) = \frac{1}{2\pi} \int_{-\infty}^{\infty} e^{-it_2 z} \varphi_Z(t_2)\, dt_2 = \frac{1}{2\pi} \int_{-\infty}^{\infty} e^{-it_2 z} [\varphi(t_2)^{n_2}]\, dt_2.$$

Thus we find

$$E(Y | Z = z) = (n_1 - n)E(X) + \frac{n}{n_2} z.$$

This can be found more easily, directly (see Exercise 320)

$$E(S_1 + S_3 | S_2 + S_3 = z) = E(S_1) + E(S_3 | S_2 + S_3 = z) = (n_1 - n)E(X) + n/z.$$

215. If $\varphi(t)$ is the characteristic function, it must be shown that for every $n \in N$ there is a characteristic function $\varphi_n(t)$ such that

$$\varphi(t) = [\varphi_n(t)]^n.$$

This is satisfied by the Pascal, Cauchy, and Gamma distributions as directly deduced from the corresponding characteristic functions in Exercises 191 and 202. For the Laplace distribution, we observe that the characteristic function is

$$\varphi(t) = (1 + t^2)^{-1} = [(1 + t^2)^{-1/n}]^n,$$

where $(1 + t^2)^{-1/n}$, *as implied by* Exercise 202, is the characteristic function of the difference of two independent and identically distributed Gamma variables with parameters $\lambda = 1$ and $s = 1/n$.

216. Let X be the length of the circumference and Y the area of the circle. We have, because of (7.1),

$$f_X(x) = \frac{1}{2\pi} f_R\left(\frac{x}{2\pi}\right) = \frac{1}{2\pi(\beta - \alpha)}, \qquad 2\pi\alpha < x < 2\pi\beta,$$

$$f_Y(y) = \frac{1}{2\sqrt{\pi y}} f_X\left(\sqrt{\frac{y}{\pi}}\right) = \frac{1}{2\sqrt{\pi y}} \frac{1}{\beta - \alpha}, \qquad \pi\alpha^2 \le y \le \pi\beta^2.$$

217. Applying (7.10) for the transformation

$$z = x + t, \qquad w = x/y,$$

with Jacobian $J(z, w) = -z/(w + 1)^2$, we find the density of z, w:

$$f(z, w) = \frac{z}{(w + 1)^2} f_{X,Y}\left(\frac{zw}{w + 1}, \frac{z}{w + 1}\right) = \frac{z}{(w + 1)^2} \lambda^2 e^{-\lambda z}, \qquad z > 0, \quad w > 0.$$

Since it can be written as a product of a function of z and a function of w, it follows that Z, W are independent; Z is Gamma and W is F with 2 and 2 degrees of freedom.

218. $X^2 - Y$ must be ≥ 0. Hence the required probability is

$$P[X^2 - Y \ge 0] = \int_{x^2 - y \ge 0} \int f(x, y) \, dx \, dy = \int_{x^2 \ge y} \int dx \, dy$$

$$= \int_0^1 dx \int_0^{x^2} dy = \frac{1}{3}.$$

Note. It is recommended that the reader show the above result by finding first the density of

$$Z = X^2 - Y,$$

that is,

$$f_Z(z) = \begin{cases} \sqrt{z + 1} & \text{for } -1 < z \le 0, \\ 1 - \sqrt{z} & \text{for } 0 < z < 1. \end{cases}$$

219. We have $\delta = v^2/g \sin 2\vartheta$ and the distribution function of δ is

$$F(x) = P\left[\frac{v^2}{g} \sin 2\vartheta \le x\right] = P\left[\sin 2\vartheta \le \frac{gx}{v^2}\right]$$

$$= P\left[\vartheta \le \frac{1}{2} \arcsin \frac{gx}{v^2}\right] + P\left[\vartheta \ge \frac{\pi}{2} - \frac{1}{2} \arcsin \frac{gx}{v^2}\right]$$

$$= \frac{2}{\pi} \arcsin \frac{gx}{v^2}.$$

Hence the density is

$$f(x) = \frac{2g}{\pi} \frac{1}{\sqrt{v^4 - g^2 x^2}}, \qquad 0 < x < \frac{v^2}{g}.$$

This is also found by using (3.2) directly.

220. (a) The density function of $Z = X/Y$ is given by

$$f_Z(z) = \int_{-\infty}^{\infty} |y| f_{(X,Y)}(yz, y)\, dy$$

$$= \frac{1}{2\pi\sigma^2 \sqrt{1-\rho^2}} \int_{-\infty}^{\infty} \exp\left[-\frac{z^2 - 2\rho z + 1}{2\sigma^2(1-\rho^2)} y^2\right] |y|\, dy$$

$$= \frac{1}{2\pi\sigma^2 \sqrt{1-\rho^2}} \int_{0}^{\infty} \exp\left[-\frac{z^2 - 2\rho z + 1}{2\sigma^2(1-\rho^2)} y^2\right] 2y\, dy$$

$$= \frac{1}{\pi} \frac{\sqrt{1-\rho^2}}{z^2 - 2\rho z + 1} \int_{0}^{\infty} e^{-t}\, dt$$

$$= \frac{1}{\pi} \frac{\sqrt{1-\rho^2}}{(z-\rho)^2 + (1-\rho^2)}, \qquad -\infty < z < \infty.$$

(b) Because of symmetry, we have

$$P[X < 0, Y > 0] = P[X > 0, Y < 0].$$

Since

$$P[XY < 0] = P[X/Y < 0] = P[X < 0, Y > 0] + P[X > 0, Y < 0],$$

it follows that

$$P[X < 0, Y > 0] = \tfrac{1}{2} P[X/Y < 0] = \tfrac{1}{2} P[Z < 0]$$

$$= \frac{1}{2} \int_{-\infty}^{0} f_Z(z)\, dz = \frac{1}{2\pi} \int_{-\infty}^{0} \frac{\sqrt{1-\rho^2}}{(z-\rho)^2 + (1-\rho^2)}\, dz$$

$$= \frac{1}{2\pi} \left[\arctan\left(\frac{z-\rho}{\sqrt{1-\rho^2}}\right)\right]_{-\infty}^{0}$$

$$= \frac{1}{2\pi}\left[\arctan\left(\frac{-\rho}{\sqrt{1-\rho^2}}\right) - \left(-\frac{\pi}{2}\right)\right]$$

$$= \frac{1}{4} - \frac{\arcsin\rho}{2\pi}.$$

221. The Cartesian coordinates (X, Y) of a point are connected with the polar coordinates (R, Θ) by the relations

$$X = R\sin\Theta, \qquad Y = R\cos\Theta.$$

The joint density of (R, Θ) is given by

$$f_{(R,\Theta)}(r, \vartheta) = f_{(X,Y)}(r\sin\vartheta, r\cos\vartheta)|J(r, \vartheta)|.$$

But

$$f_{(X,Y)}(r\sin\vartheta, r\cos\vartheta) = \frac{1}{2\pi\sigma^2}e^{-r^2/2\sigma^2}$$

and $|J(r, \vartheta)| = r$. Thus we obtain

$$f_{(R,\Theta)}(r, \vartheta) = \frac{1}{2\pi\sigma^2}e^{-r^2/2\sigma^2}\cdot r, \qquad 0 < r < \infty, \quad 0 \le \vartheta < 2\pi.$$

Hence R and Θ are independent.
The marginal density $f_R(r)$ of R is

$$f_R(r) = \int_0^{2\pi} f_{(R,\Theta)}(r, \vartheta)\,d\vartheta = \frac{r}{\sigma^2}e^{-r^2/2\sigma^2}\frac{1}{2\pi}\int_0^{2\pi}d\vartheta = \frac{r}{\sigma^2}e^{-r^2/2\sigma^2}, \qquad 0 < r < \infty,$$

that is, the Rayleigh distribution, and Θ is uniform in the interval $(0, 2\pi)$.

222. (a) If $f_Y(y)$ is the density of the random variable $Y = aX + b$ then

$$f_Y(y) = \frac{1}{|a|}f_X\left(\frac{y-b}{a}\right) = \begin{cases} 1/a & (a > 0), \quad b < y < a + b, \\ -1/a & (a < 0), \quad a + b < y < b, \end{cases}$$

that is, Y has uniform density in the interval $(b, a + b)$ if $a > 0$, or in the interval $(a + b, b)$ if $a < 0$.
(b) Let $A > 0, B > 0$; then

$$x = x(y) = -\frac{B}{2A} + \frac{\sqrt{B^2 + 4(y - C)A}}{2A} > 0 \qquad \text{for} \quad C < y < A + B + C,$$

and thus

$$f_Y(y) = f_X(x)\frac{1}{\sqrt{B^2 + 4A(y - C)}}, \qquad C < y < A + B + C.$$

Similarly, for the other cases of the constants.

223. (a) The density of $Y = X_1 + X_2$ is given by

$$f_Y(y) = \int_{-\infty}^{\infty} f_{X_1}(y - x_2) f_{X_2}(x_2)\, dx_2$$

$$= \begin{cases} \displaystyle\int_0^y dx_2 = y, & 0 \le y \le 1, \\[2mm] \displaystyle\int_{y-1}^1 dx_2 = 1 - (y - 1) = 2 - y, & 1 \le y \le 2, \end{cases}$$

that is,

$$f_Y(y) = 1 - |1 - y|, \qquad 0 \le y \le 2.$$

(b) The density of $Z = X_1 - X_2$ is given by

$$f_Z(z) = \int_{-\infty}^{+\infty} f_{X_1}(z + x_2) f_{X_2}(x_2)\, dx_2 = \begin{cases} \displaystyle\int_{-z}^1 dx_2 = 1 + z, & -1 \le z \le 0, \\[2mm] \displaystyle\int_0^{1-z} dx_2 = 1 - z, & 0 \le z \le 1, \end{cases}$$

that is,

$$f_z(z) = 1 - |z|, \qquad -1 \le z \le 1. \tag{$*$}$$

(c)

$$W = |X_1 - X_2| = |Z|.$$

We have

$$F_N(w) = P[W \le w] = P[|Z| \le w] = P[-w \le Z \le w] = F_z(w) - F_z(-w)$$

and thus

$$f_W(w) = f_Z(w) + f_Z(-w).$$

Because of $(*)$, we have

$$f_W(w) = (1 - |w|) + (1 - |-w|) = 2(1 - w), \qquad 0 \le w \le 1.$$

(d) The density of $V = X_1/X_2$ is given by (see (7.7))

$$f_V(v) = \int_{-\infty}^{\infty} |x_2| f_{X_1}(x_2 v) f_{X_2}(x_2)\, dx_2 = \begin{cases} \displaystyle\int_0^1 x_2\, dx_2 = \frac{1}{2}, & 0 \le v \le 1, \\[2mm] \displaystyle\int_0^{1/v} x_2\, dx_2 = \frac{1}{2v^2}, & v \ge 1. \end{cases}$$

224. Let X_1 be the time of arrival of A and let X_2 be the time of arrival of B, then

$$f_{X_j}(x_j) = 1, \qquad 0 \le x_j \le 1, \quad j = 1, 2,$$

$$F_{X_j}(x_j) = x_j \qquad 0 \le x_j \le 1, \quad j = 1, 2.$$

Given that the lunch lasts 0.5 hours, the probability p of meeting (event A) is given by

$$p = P(A) = P[|X_1 - X_2| \le 0.5] = P[-0.5 < X_1 - X_2 < 0.5].$$

The density of $Z = X_1 - X_2$, from Exercise 223, is given by

$$f_z(z) = 1 - |z|, \quad |z| \le 1.$$

Hence

$$p = \int_{-0.5}^{0} (1 + z)\, dz + \int_{0}^{0.5} (1 - z)\, dz = \frac{1}{2}[(1 + z)^2]_{0.5}^{0} - \frac{1}{2}[(1 - z)^2]_{0}^{0.5}$$

$$= \frac{3}{4}.$$

If T is the time of meeting then $T = \max(X_1, X_2)$ and let $f_T(t)$ denote the (unconditional) density of T. Then, taking the interval $[0, 1]$ we have

(a)

$$f_T(t) = 1/4 \quad \text{if} \quad t \notin [0, 1],$$
$$f_T(t) = 2t \quad \text{if} \quad 0 \le t \le 1/2,$$
$$f_T(t) = 1 \quad \text{if} \quad 1/2 \le t \le 1.$$

Hence, the conditional density of T given A, since $P(A) = 3/4$, is given by

$$f_T(t|A) = \frac{f_T(t)}{p} = \frac{4}{3} f_T(t) \quad \text{for} \quad 0 \le t \le 1. \tag{1}$$

(b) Similarly as in (1).

225. The density of the product $Z = XY$ is given by the formula

$$f_Z(z) = \int_{-\infty}^{\infty} \frac{1}{|y|} f_Y(y) f_X\left(\frac{z}{y}\right) dy = \frac{1}{\pi\sigma^2} \int_{|z|}^{\infty} \frac{1}{|y|} y e^{-y^2/2\sigma^2} \frac{1}{\sqrt{1 - z^2/y^2}}\, dy.$$

For the limits of integration of y, we have $|z| < y < \infty$, because $|X| < 1$ and $y > 0$. Thus

$$f_Z(z) = \frac{1}{\pi\sigma^2} \int_{|z|}^{\infty} y e^{-y^2/2\sigma^2} \frac{1}{|y|} \frac{1}{\sqrt{1 - z^2/y^2}}\, dy$$

$$= \frac{1}{\pi\sigma} e^{-z^2/2\sigma^2} \int_{0}^{\infty} e^{-t^2/2}\, dt = \frac{1}{\pi\sigma} e^{-z^2/2\sigma^2} \frac{1}{2}\sqrt{2\pi}$$

$$= \frac{1}{\sigma\sqrt{2\pi}} e^{-z^2/2\sigma^2}, \quad -\infty < z < +\infty,$$

that is, the density of $N(0, \sigma^2)$.

226. (a)

$$Y = \sum_{j=1}^{n} 2 \log X_j = \sum_{j=1}^{n} Y_j$$

where

$$Y_j = -2 \log X_j, \qquad j = 1, 2, \ldots, n,$$

independent variables with density

$$f_{Y_j}(y_j) = f_{X_j}(e^{-y_j/2}) \left| \frac{d}{dy_j} e^{-y_j/2} \right| = \frac{1}{2} e^{-y_j/2}, \qquad y_j > 0.$$

Thus Y_j has the Gamma distribution with parameters $\lambda = 1/2$ and $s = 1$, Exercise 203 (i.e., exponential distribution). But the Gamma distribution has the reproductive property and hence Y, as the sum of n independent and isonomic Gamma variables, has Gamma distribution with parameters $\lambda = 1/2$ and $s = n$, that is,

$$f_Y(y) = \frac{1}{2^n \Gamma(n)} y^{n-1} e^{-y/2}, \qquad y > 0,$$

that is, Y has the χ^2 distribution with $2n$ degrees of freedom.

(b) The joint density of the random variables (see (7.11))

$$\xi = \sqrt{-2 \log X_1} \cos 2\pi X_2, \qquad \eta = \sqrt{-2 \log X_1} \sin 2\pi X_2$$

is given by

$$f(\xi, \eta) = f_{(X_1, X_2)}(x_1(\xi, \eta), x_2(\xi, \eta)) |J(\xi, \eta)|.$$

But

$$J(\xi, \eta) = \frac{1}{2\pi} e^{-(\xi^2 + \eta^2)/2}$$

and hence

$$f(\xi, \eta) = \frac{1}{\sqrt{2\pi}} e^{-\xi^2/2} \frac{1}{\sqrt{2\pi}} e^{-\eta^2/2}, \qquad -\infty < \xi < \infty, \quad -\infty < \eta < \infty.$$

From the above relation, we deduce that ξ, η are independent $N(0, 1)$.

227. Let $Y = F(X)$. Then

$$F_Y(y) = P[Y \le y] = P[F(X) \le y]$$

$$= \begin{cases} 0, & y \le 0, \\ P[X \le F^{-1}(y)] = F[F^{-1}(y)] = y, & 0 \le y \le 1, \\ 1, & y \ge 1. \end{cases} \qquad (1)$$

If $X_1, X_2, \ldots, X_n$ are pseudorandom numbers, then $Y_1 = F^{-1}(X_1), \ldots, Y_n = F^{-1}(X_n)$ are independent, and by (1) they are isonomic (identically distributed) with distribution function F.

228. By Exercise 239 we have

$$P[X \in \delta] = R,$$

where R has density given by (2) of Exercise 239 with $n = 10$. Then the required probability is

$$P[R \geq 0.95] = 10 \times 9 \int_{0.95}^{1} r^8(1 - r) \, dr = 1 - 10(0.95)^9 + 9(0.95)^{10}.$$

229. (a)

$$f_X(x) = f_Y(\log x) \left| \frac{d}{dx}(\log x) \right| = \frac{1}{\sigma\sqrt{2\pi}x} \exp\left[-\frac{1}{2\sigma^2}(\log x - \mu)^2 \right], \qquad x > 0.$$

(b) We have

$$E(X) = E[e^Y] = M_Y(1),$$

$$\mathrm{Var}(X) = E(X^2) - [E(X)]^2 = M_Y(2) - [M_Y(1)]^2,$$

where $M_Y(t) = e^{\mu t + \sigma^2 t^2/2}$ is the moment generating functions of the normal random variable Y. Hence

$$E(X) = e^{\mu + \sigma^2/2} \quad \text{and} \quad \mathrm{Var}(X) = e^{2\mu + 2\sigma^2} - e^{2\mu + \sigma^2}.$$

(c) $\log X_i$ is normal; hence the sum $\log X_1 + \cdots + \log X_n = \log(\prod_{i=1}^{n} X_i)$ is normal. Hence it follows that $Y = \prod_{i=1}^{n} X_i$ is lognormal.

230. From the definition of the distribution function we have

$$F_Z(z) = P[Z \leq z] = P[X \leq z, Y \leq z] = F(z, z),$$

$$F_W(w) = P[W \leq w] = 1 - P[W > w] = 1 - P[X > w, Y > w].$$

Using the addition theorem

$$P[A \cup B] = P(A) + P(B) - P(AB) = 1 - P(A'B')$$

with $A = \{X \leq w\}$, $B = \{Y \leq w\}$, we have

$$1 - P[X > w, Y > w] = P(X \leq w) + P(Y \leq w) - P[X \leq w, Y \leq w]$$

$$= F_X(w) + F_Y(w) - F(w, w).$$

If F is continuous, we obtain the densities of Z, W by taking the derivative of the corresponding distribution function; thus, we have

$$f_Z(z) = \frac{d}{dz}F(z, z) = \frac{d}{dz} \int_{-\infty}^{z} \int_{-\infty}^{z} f(x, y) \, dx \, dy$$

$$= \int_{-\infty}^{z} f(x, z) \, dx + \int_{-\infty}^{z} f(z, y) \, dy, \qquad (1)$$

$$f_W(w) = f_X(w) + f_Y(w) - \int_{-\infty}^{w} f(x, w) \, dx - \int_{-\infty}^{w} f(w, y) \, dy.$$

231. Because of the independence of X, Y (see Exercise 230(1)) we have

$$f_Z(z) = 2\Phi(z)\varphi(z),$$

where Φ and φ is the distribution function and the density of $N(0, 1)$, respectively. Hence

$$E(Z) = \int_{-\infty}^{\infty} z f_Z(z)\, dz = \frac{1}{\pi} \int_{-\infty}^{\infty} z e^{-z^2/2}\, dz \int_{-\infty}^{z} e^{-x^2/2}\, dx$$

$$= \frac{1}{\pi} \int_{-\infty}^{\infty} e^{-x^2/2}\, dx \int_{-x}^{\infty} z e^{-z^2/2}\, dz = \frac{1}{\pi} \int_{-\infty}^{\infty} e^{-x^2}\, dx = \frac{1}{\sqrt{\pi}}.$$

232. The joint density of $Y_1 = X_1$, $Y_2 = X_1 + X_2$, ..., $Y_k = X_1 + X_2 + \cdots + X_k$ is given by

$$g(y_1, y_2, \ldots, y_k) = f(y_1, y_2 - y_1, \ldots, y_k - y_{k-1})|J(y_1, y_2, \ldots, y_k)|.$$

Since $X_1 = Y_1$, $X_2 = Y_2 - Y_1$, ..., $X_k = Y_k - Y_{k-1}$, we have $J(y_1, y_2, \ldots, y_k) = 1$ and so

$$g(y_1, \ldots, y_k) = \frac{\Gamma(n_1 + n_2 + \cdots + n_{k+1})}{\Gamma(n_1)\Gamma(n_2)\ldots\Gamma(n_{k+1})} y_1^{n_1-1}(y_2 - y_1)^{n_2-1}\ldots(y_k - y_{k-1})^{n_k-1}$$

$$\cdot(1 - y_k)^{n_{k+1}-1}, \qquad 0 \le y_1 \le y_2 \le \cdots \le y_k \le 1.$$

Then the marginal density of Y_k is

$$g_k(y_k) = \int_0^{y_2} \int_0^{y_3} \cdots \int_0^{y_k} g(y_1, y_2, \ldots, y_k)\, dy_1\, dy_2 \ldots dy_{k-1}.$$

But

$$\int_0^{y_2} y_1^{n_1-1}(y_2 - y_1)^{n_2-1}\, dy_1 = y_2^{n_1+n_2-1} \int_0^1 w^{n_1-1}(1 - w)^{n_2-1}\, dw$$

$$= y^{n_1+n_2-1} \frac{\Gamma(n_1)\Gamma(n_2)}{\Gamma(n_1 + n_2)}.$$

Thus, by successive integrations, we finally obtain

$$g_k(y_k) = \frac{\Gamma(n_1 + n_2 + \cdots + n_{k+1})}{\Gamma(n_1 + n_2 + \cdots + n_k)\Gamma(n_{k+1})} y_k^{n_1+\cdots+n_k-1}(1 - y_k)^{n_{k+1}-1}$$

$$= \frac{1}{B(n_1 + n_2 + \cdots + n_k, n_{k+1})} y_k^{n_1+n_2+\cdots+n_k-1}(1 - y_k)^{n_{k+1}-1}, \quad 0 \le y_k \le 1.$$

233. The joint density of Y_1, Y_2, ..., Y_k, $Y = X_1 + X_2 + \cdots + X_{k+1}$, is given by

$$f_1(yy_1)f_2(yy_2)\ldots f_k(yy_k)f_{k+1}(y(1 - y_1 - y_2 - \cdots - y_k))|J(y_1, \ldots, y_k, y)|,$$

$$\tag{1}$$

where

$$f_i(x_i) = \frac{\lambda^{n_i}}{\Gamma(n_i)} x_i^{n_i-1} e^{-\lambda x_i}, \qquad x_i > 0, \quad i = 1, 2, \ldots, k+1,$$

and since $x_1 = yy_1$, $x_2 = yy_2$, $\ldots$, $x_k = yy_k$, $x_{k+1} = y(1 - y_1 - \cdots - y_k)$, we have

$$J(y_1, \ldots, y_k, y) = \begin{vmatrix} y & 0 & \cdots & 0 & y_1 \\ 0 & y & \cdots & 0 & y_2 \\ \cdots\cdots\cdots\cdots\cdots\cdots\cdots\cdots\cdots\cdots\cdots\cdots\cdots\cdots\cdots \\ 0 & 0 & \cdots & y & y_k \\ -y & -y & \cdots & -y & (1 - y_1 - \cdots - y_k) \end{vmatrix} = y^k.$$

Then (1) is written as

$$\frac{\lambda^{n_1+n_2+\cdots+n_{k+1}}}{\Gamma(n_1)\Gamma(n_2)\ldots\Gamma(n_{k+1})} y_1^{n_1-1} \ldots y_k^{n_k-1} (1 - y_1 - \cdots - y_k)^{n_{k+1}-1}$$

$$\cdot y^{n_1+n_2+\cdots+n_{k+1}-1} e^{-\lambda y},$$

which gives the required Dirichlet density.

234. (a) The density of the random variable $Z = X + Y$ is given by

$$f_Z(z_k) = \sum_{y_j=0}^{z_k} f_X(z_k - y_j) f_Y(y_j) = \sum_{y_j=0}^{z_k} \frac{e^{-\lambda} \lambda^{z_k - y_j}}{(z_k - y_j)!} \frac{e^{-\mu} \mu^{y_j}}{y_j!}$$

$$= e^{-(\lambda+\mu)} \frac{1}{z_k!} \sum_{y_j=0}^{z_k} \binom{z_k}{y_j} \mu^{y_j} \lambda^{z_k - y_j}$$

$$= e^{-(\lambda+\mu)} \frac{(\lambda + \mu)^{z_k}}{z_k!}, \qquad z_k = 0, 1, 2, \ldots,$$

that is, Poisson with parameter $\lambda + \mu$.

(b)

$$P[X = k | X + Y = n] = \frac{P[X = k, Y = n - k]}{P[Z = n]}$$

$$= \frac{e^{-\lambda} \dfrac{\lambda^k}{k!} e^{-\mu} \dfrac{\mu^{n-k}}{(n - k)!}}{e^{-(\lambda+\mu)} \dfrac{(\lambda + \mu)^n}{n!}}$$

$$= \binom{n}{k} \left(\frac{\lambda}{\lambda + \mu}\right)^k \left(\frac{\mu}{\lambda + \mu}\right)^{n-k}, \qquad k = 0, 1, \ldots, n,$$

that is, binomial with parameters n and $p = \lambda/(\lambda + \mu)$.

235. (a)

$$P[X + Y = n] = \sum_{k=0}^{n} P[X = k]P[Y = n - k]$$

$$= \sum_{k=0}^{n} \binom{\Lambda}{k} p^k q^{\Lambda - k} \binom{M}{n - k} p^{n-k} q^{M-(n-k)}$$

$$= p^n q^{\Lambda + M - n} \sum_{k=0}^{n} \binom{\Lambda}{k}\binom{M}{n - k}$$

$$= \binom{\Lambda + M}{n} p^n q^{\Lambda + M - n}, \qquad n = 0, 1, \ldots, M + \Lambda.$$

(b)

$$P[X = k \,|\, X + Y = n] = \frac{P[X = k, Y = n - k]}{P[X + Y = n]} = \frac{P[X = k]P[Y = n - k]}{P[X + Y = n]}$$

$$= \frac{\binom{\Lambda}{k} p^k q^{\Lambda - k} \binom{M}{n - k} p^{n-k} q^{M-n+k}}{\binom{\Lambda + M}{n} p^n q^{\Lambda + M - n}} = \frac{\binom{\Lambda}{k}\binom{M}{n - k}}{\binom{\Lambda + M}{n}}.$$

236. We have

$$(n - 1)s^2 = \sum_{i=1}^{n} X_i^2 - n\bar{X}^2 = \sum_{i=1}^{n} Y_i^2 - (\sqrt{n}\bar{X})^2 = \sum_{i=1}^{n} Y_i^2 - Y_n^2 = \sum_{i=1}^{n-1} Y_i^2, \quad (1)$$

since the transformation $y = Hx$ gives

$$y'y = \sum_{j=1}^{n} y_j^2 = x'H'Hx = x'x = \sum_{i=1}^{n} x_i^2.$$

The covariance matrix of $Y = (Y_1, Y_2, \ldots, Y_n)'$ is $D(Y) = HD(X)H' = \sigma^2 HIH'$ $= \sigma^2 I$, that is, $Y_1, \ldots, Y_n$ are independent $N(0, \sigma^2)$. Hence $Y_n = \sqrt{n}\bar{X}, \bar{X}$ is also independent of $Y_1, \ldots, Y_{n-1}$, and therefore of s^2, because of (1). Moreover, the distribution of $(n - 1)s^2/\sigma^2$ is χ_{n-1}^2. The independence of $\bar{X}$ and s^2 is also deduced from the fact that

$$\text{Cov}(X_i - \bar{X}, \bar{X}) = 0, \qquad i = 1, 2, \ldots, n.$$

This, by the joint normality of $\bar{X}, X_i - \bar{X}$, implies their independence, that is, $\bar{X}$ is independent of $X_1 - \bar{X}, X_2 - \bar{X}, \ldots, X_n - \bar{X}$; and hence also of s^2. For the $\text{Cov}(\bar{X}, s^2)$, taking $E(X_i) = 0$, without any loss of generality, we have by (1)

$$\text{Cov}(\bar{X}, s^2) = \frac{1}{n - 1} E\left[\bar{X} \sum_{i=1}^{n} X_i^2\right] - \frac{n}{n - 1} E(\bar{X}^3)$$

$$= \frac{1}{n(n - 1)} E\left[\sum_{i=1}^{n} X_i \cdot \sum_{i=1}^{n} X_i^2\right] - \frac{1}{n(n - 1)} E\left[\sum_{i=1}^{n} X_i^s\right]$$

$$= \frac{1}{n - 1}\mu_3 - \frac{1}{n(n - 1)}\mu_3 = \frac{1}{n}\mu_3.$$

237. Show that

$$\bar{x} - \bar{y} = \frac{1}{m} \sum_{i=1}^{m} v_j = \bar{v},$$

where

$$v_j = u_j + \frac{1}{\sqrt{mn}} \sum_{i=1}^{m} y_i - \bar{y}$$

are independent $N(\mu_1 - \mu_2, \sigma^{*2})$ with

$$\sigma^{*2} = \mathrm{Var}(u_j) = \sigma_1^2 + \frac{m}{n}\sigma_2^2 \qquad \text{and} \qquad \sum_{i=1}^{m}(u_j - \bar{u})^2 = \sum_{i=1}^{m}(v_i - \bar{v})^2.$$

Hence for $\mu_1 = \mu_2$ we have the required distribution.

238. The density of the random sample $X_1, X_2, \ldots, X_n$ from a continuous distribution with density f is

$$f(x_1, \ldots, x_n) = f(x_1) \ldots f(x_n).$$

Given that the $n!$ permutations of $X_1, \ldots, X_n$ give the same ordered sample $X_{(1)}, X_{(2)}, \ldots, X_{(n)}$, it follows that the required density is

$$f^*(x_1, x_2, \ldots, x_n) = n!\, f(x_1, x_2, \ldots, x_n) = n! \prod_{i=1}^{n} f(x_i).$$

239. (a) By Exercise 227, $Y = \min(Y_1, \ldots, Y_n)$, $Z = \max(Y_1, \ldots, Y_n)$, where $Y_1, Y_2, \ldots, Y_n$ constitute a random sample from the uniform in $(0, 1)$. Working as in Exercise 238, we find that the joint density of Y, Z is

$$f(y, z) = n(n - 1)(z - y)^{n-2}, \qquad 0 < y < z. \tag{1}$$

(b) The density of $R = Z - Y$ is easily seen to be

$$f_R(r) = n(n - 1) \int_0^{1-r} r^{n-2}\, dy = n(n - 1)r^{n-2}(1 - r), \qquad 0 < r < 1. \tag{2}$$

Hence we have

$$E(R) = n(n - 1) \int_0^1 r^{n-1}(1 - r)\, dr = \frac{n - 1}{n + 1}.$$

240. The joint density of $X_{(j)}, X_{(k)}$ $(j < k)$, is given by

$$f(x_j, x_k) = \frac{n!}{(j - 1)!\,(k - j - 1)!\,(n - k)!}[F(x_j)]^{j-2}[F(x_k) - F(x_j)]^{k-j-1}$$

$$\cdot [1 - F(x_k)]^{n-k}f(x_j)f(x_k), \qquad x_j < x_k,$$

because $j - 1$ of the X_i must be less than x_j, $k - j - 1$ between X_j and X_k, $n - k$ must be larger than x_k, one in the interval $(x_j, x_j + dx_j)$ and one in the interval $(x_k, x_k + dx_k)$ ($f(x)$ and $F(x)$ represent the density and the distribution function of the original distribution, respectively). Hence we have

$E[X_{(k+1)} - X_{(k)}]$

$$= \int_{-\infty}^{\infty} \int_{-\infty}^{\infty} (y - x)f(x, y) \, dx \, dy$$

$$= \frac{n!}{(k-1)! \, (n-k-1)!} \left\{ \int_{-\infty}^{\infty} F^{k-1}(x)f(x) \, dx \int_{x}^{\infty} y[1 - F(y)]^{n-k-1} f(y) \, dy \right.$$

$$\left. - \int_{-\infty}^{\infty} xF^{k-1}(x)f(x) \, dx \int_{x}^{\infty} [1 - F(y)]^{n-k-1} f(y) \, dy \right\}.$$

Integrating by parts gives the result. This can also be found from the relation

$$E[X_{(k+1)} - X_{(k)}] = E[X_{(k+1)}] - E[X_{(k)}],$$

using the density of $X_{(k)}$ in Exercise 241.

241. (a) The density of $X_{(k)}$ is (cf. Exercise 240)

$$f_k(x) = \frac{n!}{(k-1)! \, (n-k)!} [F(x)]^{k-1} [1 - F(x)]^{n-k} f(x), \tag{1}$$

where for the uniform $f(x) = 1 \ (0 \le x \le 1)$, $F(x) = x \ (0 \le x \le 1)$. Hence we have

$$E[X_{(k)}] = \frac{n!}{(k-1)! \, (n-k)!} \int_{0}^{1} x^k (1-x)^{n-k} \, dx = \frac{k}{n+1}.$$

(b) The density of R is

$$f_R(r) = n(n-1) \int_{-\infty}^{\infty} f(x)f(x+r)[F(x+r) - F(x)]^{n-2} \, dx. \tag{2}$$

For the uniform density we have

$$f_R(r) = n(n-1) \int_{0}^{1-r} r^{n-2} \, dx = n(n-1)r^{n-2}(1-r). \tag{3}$$

(c) From (3) we find

$$E(R) = n(n-1) \int_{0}^{1} r^{n-1}(1-r) \, dr = \frac{n-1}{n+1}.$$

242. Because of (1) of Exercise 241, the density of Y_2 is

$$f_2(x) = 3! \, (x - \vartheta + \tfrac{1}{2})(\vartheta - x + \tfrac{1}{2}), \qquad \vartheta - \tfrac{1}{2} < x < \vartheta + \tfrac{1}{2}.$$

Hence the required probability is

$$6 \int_{\vartheta-0.4}^{\vartheta+0.4} (x - \vartheta + \tfrac{1}{2})(\vartheta - x + \tfrac{1}{2}) \, dx = 6 \int_{0.1}^{0.9} y(1 - y) \, dy = 0.944.$$

243. The joint density of Y_1, Y_2, Y_3, by Exercise 238, is

$$g(y_1, y_2, y_3) = 3! \, (2y_1)(2y_2)(2y_3) = 48 y_1 y_2 y_3.$$

According to (7.11) the density of Z_1, Z_2, Z_3 is

$$f(z_1, z_2, z_3) = g(y_1, y_2, y_3)|J(z_1, z_2, z_3)| = 48 z_1 z_2^3 z_3^5,$$

because

$$J(z_1, z_2, z_3) = \left| \left(\frac{\partial y_i}{\partial z_j} \right) \right| = z_2 z_3^2.$$

So we have the independence of Z_1, Z_2, Z_3.

244. Show that the density of $Y = \sqrt{X_1 X_2}$ is given by

$$f_Y(y) = 2y \int_{-\infty}^{\infty} \frac{1}{|x|} f_{X_1}\left(\frac{y^2}{x}\right) f_{X_2}(x) \, dx$$

$$= c y^{2n_1} \int_{y^2}^{1} x^{-3/2} \left[\left(1 - \frac{y^2}{x}\right)(1 - x) \right]^{n_2 - 1} dx.$$

245. Direct application of (7.11) with $z = g_1(x, y) = y - x$, $w = g_2(x, y) = x$. Z is Gamma with parameter n and it is independent of X.

246. Apply Exercise 244.

247. Application of (7.6). Y is Gamma with parameter $m + n$.

248. Setting $X = R \cos \theta$, $Y = R \sin \theta$, we have

$$Z = R \sin 2\theta, \qquad W = R \cos 2\theta, \tag{1}$$

where, by Exercise 221, R and θ are independent with densities

$$f_R(r) = \frac{r}{\sigma^2} e^{-r^2/2\sigma^2}, \quad r > 0, \qquad f(\theta) = \frac{1}{2\pi}, \quad 0 < \theta < 2\pi.$$

Now the joint density of Z, W can easily be found to be

$$f(z, w) = \frac{1}{2\pi\sigma^2} \exp\left(-\frac{z^2 + w^2}{2\sigma^2}\right),$$

which shows the independence of Z and W.

249. By applying (7.12) to (5.10) we find that Y is $N(A\mu, A\Sigma A')$.

250. Let (X_i, Y_i) be the impact point of A_i ($i = 1, 2$). The required probability is equal to

$$P[R_1 < R_2] = P[R_1^2 < R_2^2] = \int_0^{\infty} F_1(r) f_2(r) \, dr = 1 - \frac{\sigma_1^2}{\sigma_1^2 + \sigma_2^2},$$

since $R_i^2 = X_1^2 + Y_1^2$, by Exercise 221, has density

$$f_i(r) = \frac{1}{2\sigma_i^2} \exp\left(-\frac{1}{2}\frac{r}{\sigma_i^2}\right),$$

and distribution function

$$F_i(r) = 1 - \exp\left(-\frac{1}{2}\frac{r}{\sigma_i^2}\right), \qquad i = 1, 2.$$

251. Apply (8.9)

252. Follows from (8.5) and

$$\text{Var}(\bar{X}_n) = \frac{1}{n^2}\left\{\sum_{i=1}^{n} \text{Var}(X_i) + 2\sum_{k=1}^{n-1} \text{Cov}(X_k, X_{k+1})\right\}$$

$$\leq \frac{1}{n^2}\{nM + 2(n-1)M\} \xrightarrow[n\to\infty]{} 0.$$

253. Follows from

$$\text{Var}(\bar{X}_n) < n^{-2}\sum_{i=1}^{n} \text{Var}(X_i) \leq n^{-1}c \to 0 \qquad \text{as} \quad n \to \infty.$$

254. Show that

$$\text{Var}(\bar{X}_n) \to 0 \qquad \text{as} \quad n \to \infty.$$

255. We will show that for every continuity point x of F_X

$$\lim_{n\to\infty} P[X_n + Y_n \leq x] = F_X(x - c). \tag{1}$$

We have

$$P[X_n + Y_n \leq x] = P[X_n + Y_n \leq x, |Y_n - c| \leq \varepsilon]$$
$$+ P[X_n + Y_n \leq x, |Y_n - c| > \varepsilon],$$

where

$$P[X_n + Y_n \leq x, |Y_n - c| > \varepsilon] \leq P[|Y_n - c| > \varepsilon] \to 0 \qquad \text{as} \quad n \to \infty,$$

and the probability tends to $F(x - c)$.

$$P[X_n + Y_n \leq x, |Y_n - c| \leq \varepsilon]$$

Similarly, one can show (b) and (c).

256. Let

$$L_n = \sum_{j=1}^{k} c_j X_{jn} \qquad \text{and} \qquad L = \sum_{j=1}^{k} c_j X_j.$$

Then

$$\varphi_{L_n}(u) = E(e^{iuL_n}) = \varphi_{\xi_n}(uc_1, uc_2, \ldots, uc_k), \tag{1}$$

$$\varphi_L(u) = E(e^{iuL}) = \varphi_\xi(uc_1, uc_2, \ldots, uc_k). \tag{2}$$

Since

$$L_n \xrightarrow{L} L \implies \varphi_{L_n}(u) \to \varphi_L(u),$$

we have

$$\varphi_{L_n}(1) \to \varphi_L(1),$$

that is,

$$\varphi_{\xi_n}(c_1, c_2, \ldots, c_k) \to \varphi_\xi(c_1, c_2, \ldots, c_k)$$

for every $(c_1, \ldots, c_k)$. Hence by the continuity theorem of characteristic functions for multivariate distributions

$$\xi_n \xrightarrow{L} \xi.$$

257. From $P[|X_k| \le M] = 1$ it follows that $X_k - E(X_k)$ are uniformly bounded and since $s_n^2 \to \infty$ for every $\varepsilon > 0$, there is an N such that for $n > N$ the relation

$$P[|X_k - E(X_k)| < \varepsilon s_n, k = 1, 2, \ldots, n] = 1$$

holds. Hence the condition of Lindeberg–Feller's theorem is satisfied and the CLT holds.

258. If X_n denotes a Poisson random variable with parameter $\lambda = n = E(X_n)$, then we have

$$e^{-n} \sum_{k=1}^{n} \frac{n^k}{k!} = P[X_n \le n] = P\left[\frac{X_n - n}{\sqrt{n}} \le 0\right] \to \Phi(0) = \frac{1}{2} \qquad \text{as} \quad n \to \infty$$

by the normal approximation to the Poisson.

259. We have

$$X \xrightarrow{P} \lambda_1, \qquad Y \xrightarrow{P} \lambda_2, \qquad X + Y \xrightarrow{L} \lambda_1 + \lambda_2,$$

$$X + Y \xrightarrow{L} N(\lambda_1 + \lambda_2, \lambda_1 + \lambda_2).$$

The result follows by application of Exercise 255(c) with $c = \sqrt{\lambda_1 + \lambda_2}$.

260. Verify that the Lindeberg–Feller condition is satisfied.

261. According to the CLT

$$P[-1 < S_n^* < 1] \to \Phi(1) - \Phi(-1) = 0.68,$$

while Chebyshev's inequality gives no information since

$$P[|S_n| \ge \delta] \le \delta^{-2} = 1 \qquad \text{for} \quad \delta = 1.$$

For $k = 2$, Chebyshev's inequality gives

$$P[-2 < S_n^* < 2] \ge 0.75,$$

while the CLT gives as a limit $\Phi(2) - \Phi(-2) = 0.9544$. Similarly, for $k = 3$, we have 0.8888 and 0.9974, respectively.

262.
$$\mu = E(X) = \int_2^\infty x24x^{-4}\,dx = 24\int_2^\infty x^{-3}\,dx = 3,$$

$$q(\delta) = P[|X - \mu| > \delta] = P[|X - 3| > \delta]$$
$$= 1 - P[|X - 3| < \delta]$$
$$= 1 - P[3 - \delta < X < \delta + 3] = 1 - \int_{3-\delta}^{3+\delta} 24x^{-4}\,dx$$
$$= 1 - 8\left[\frac{1}{(3-\delta)^3} - \frac{1}{(3+\delta)^3}\right]$$
$$= \begin{cases} 8(3+\delta)^{-3}, & \delta \geq 1, \\ 1 - 8[(3-\delta)^{-3} - (3+\delta)^{-3}], & \delta < 1. \end{cases}$$

According to Chebyshev's inequality

$$P[|X - \mu| > \delta] \leq \frac{\text{Var}(X)}{\delta^2} = \frac{3}{\delta^2}.$$

The following table and figure show the difference between $q(\delta)$ and $3\delta^{-2}$. Note that Chebyshev's inequality for $\delta \leq \sigma = \sqrt{3}$ gives the obvious upper bound 1.

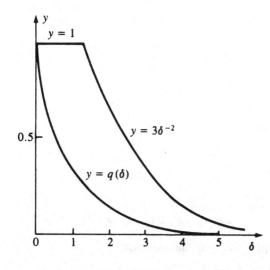

δ	$q(\delta)$	By Chebyshev
1/2	0.673	1
1	0.125	1
2	0.064	0.750
3	0.037	0.333
5	0.016	0.120

263. Applying the CLT we find that n will satisfy the relation

(a) $P[|\bar{X} - \mu| > 0.05\sigma] \approx P[|Z| > 0.05\sqrt{n}] < 0.05.$

Hence $0.05\sqrt{n} > 1.96$ ($Z \sim N(0, 1)$). Thus $n \geq 1537$

(b)
$$P[|\bar{X} - \mu| > 0.05\mu] \leq P\left[\frac{|\bar{X} - \mu|}{\sigma/\sqrt{n}} > \frac{0.05 \times 5}{\sigma/\sqrt{n}}\right]$$

$$\approx P\left[|Z| > \frac{0.25}{3}\sqrt{n}\right].$$

Hence $\sqrt{n} > 196 \cdot (3/25)$ and $n > 553$.

According to Chebyshev's inequality n must satisfy the relation

(a)
$$P[|\bar{X} - \mu| > 0.5\sigma] \leq \frac{1}{(0.05)^2 n} < 0.05 \Rightarrow n > 20^3,$$

(b)
$$P[|\bar{X} - \mu| > 0.05\mu] \leq \frac{\sigma^2}{n(0.05)^2\mu^2} < \frac{9}{(0.25)^2 n}$$

$$< 0.05 \Rightarrow n > 2880.$$

264. (i) According to the De Moivre–Laplace theorem and putting $Z \sim N(0, 1)$, the required size n of the sample will satisfy the relation

$$P\left[\left|\frac{S_n}{n} - p\right| < 0.01\right] \approx P\left[|Z| < \frac{\sqrt{n}}{100\sqrt{pq}}\right] > 0.95,$$

from which

$$n > 196^2 pq.$$

Then we have (a) $n > 196^2 \times 0.3 \times 0.7$ or $n > 8067$. (b) The maximum value of pq is attained for $p = 0.5$. Hence $n > 196^2 \times 0.25$ or $n > 9604$.

(ii) Working as in (i), we find (a) $n > 323$, (b) $n > 384$.

265. According to the Moivre–Laplace theorem and putting $Z \sim N(0, 1)$, the number of seats k must satisfy the relation

$$P[S_{300} > k] \approx P\left[Z > \frac{k - 300 \times 1/3}{\sqrt{300 \times 1/3 \times 2/3}}\right] < \frac{1}{20},$$

$$\frac{k - 100}{\sqrt{300/3}} > 1.645 \quad \text{or} \quad k \geq 112.$$

266. If the X_i's are independent uniform random variables in $(-0.05, 0.05)$, we have, according to the CLT with $\mu = 0$, $\text{Var}(X_i) = 0.01/12$,

$$P\left[\left|\sum_{i=1}^{1000} X_i\right| < 2\right] \approx P\left[|z| < \frac{2}{0.91}\right] = P[|Z| < 2.19] = 2\Phi(2.19) - 1$$

$$= 0.97.$$

267. As in Exercise 266 we have, since $E(X_i) = 5$, $\text{Var}(X_i) = 90^2/12$,

$$P\left[\sum_{i=1}^{60} X_i > 500\right] \approx P\left[Z > \frac{200}{220}\right] \approx 1 - \Phi(0.91) = 0.18.$$

(b) c must satisfy the relation

$$P\left[Z < \frac{c - 300}{201.6}\right] \le \frac{1}{10} \Rightarrow \frac{c - 300}{201.6} = -1.28 \Rightarrow c = 42.$$

268. The profit X_i of the casino during the game i, is a two-valued random variable X_i with $E(X_i) = -1/37$. Thus

$$P\left[\sum_{k=1}^{n} X_i \ge 1,000\right] \approx P\left[Z \ge \frac{1,000 - n/37}{\sqrt{n\sigma}}\right] \ge \frac{1}{2},$$

where Z is $N(0, 1)$, and hence

$$1,000 - \frac{n}{37} \le 0 \Rightarrow n \ge 37,000,$$

$$P[\text{of a loss}] = P\left[\sum_{i=1}^{37,000} X_i < 0\right] \approx P\left[Z < -\frac{1,000}{\sqrt{37,000}}\right] \approx 0.$$

269. Use Exercise 215(d) and the CLT, or show directly that the characteristic function of the standardized Gamma tends to $\exp(-t^2/2)$.

270. $$E[g(X)] = \int_{-\infty}^{\infty} g(x)\, dF(x) \ge \int_{a}^{b} g(x)\, dF(x) \ge c \int_{a}^{b} dF(X)$$

$$= cP[X \in (a, b)].$$

271. Take as g of Exercise 270, $g(x) = (x + c)^2$ with $c > 0$. If $E(X) = 0$ and $E(X^2) = \sigma^2$, then the minimum of $(t + c)^{-2} \cdot E(X + c)^2$ is obtained when $c = \sigma^2/t$. Then we have the required inequality.

272. From the event $\{|X_1 - X_2| \ge t\}$ we have that at least one of $\left\{|X_1| > \frac{1}{2} > t\right\}, \left\{|X_2| > \frac{1}{2}t\right\}$ occurs. Thus we have

$$P[|X_1 - X_2| \ge t] \le P[|X_1| > \tfrac{1}{2}t \text{ or } |X_2| \ge \tfrac{1}{2}t] \le 2P[|X_1| > \tfrac{1}{2}t].$$

273. See Exercise 271.

274. If the function $g(y)$ is decreasing for $y > 0$, then for $\lambda^2 > 0$ we have

$$\lambda^2 \int_{\lambda}^{\infty} g(y)\, dy \le \frac{4}{9} \int_{0}^{\infty} y^2 g(y)\, dy. \tag{1}$$

(i) In the special case when g is constant

$$g(y) = \begin{cases} A & \text{for } 0 < y < c, \\ 0 & \text{for } y \ge c, \end{cases}$$

we have

$$\lambda^2 \int_{\lambda}^{c} g(y)\, dy = A\lambda^2(c - \lambda).$$

This is maximized for $\lambda = (2/3)c$, as we can easily see by differentiation.

(ii) We define the function

$$h(y) = \begin{cases} g(\lambda) = \text{const.}, & 0 < y < \lambda + a, \\ 0 & y \geq \lambda + a, \end{cases}$$

where $ag(\lambda) = \int_\lambda^\infty g(y)\, dy$. Then we have

$$\lambda^2 \int_\lambda^\infty g(y)\, dy = \lambda^2 \int_\lambda^\infty h(y)\, dy \leq \frac{4}{9} \int_\lambda^\infty y^2 h(y)\, dy \leq \frac{4}{9} \int_0^\infty y^2 g(y)\, dy.$$

Let $f(x)$ be the density of the continuous random variable X. Since x_0 is the only mode (maximum) of f, it is decreasing for $x > x_0$. Hence taking

$$g(|X^*|) = \tau f(x_0 + \tau |X^*|), \qquad \text{with} \quad X^* = \frac{X - x_0}{\tau},$$

we have by (1) that

$$P[|X - x_0| \geq \lambda \tau] \leq \frac{4}{9} \frac{E\left[\dfrac{X - x_0}{\tau}\right]^2}{\lambda^2} = \frac{4}{9\lambda^2}.$$

275. Application of Markov's inequality and of the relation

$$P[|X| \geq c] = P[g(X) \geq g(c)].$$

276.
$$\begin{aligned} E[g(X)] &= E[g(X)||X| \leq c]P[|X| \leq c] \\ &\quad + E[g(X)||X| > c]P[|X| > c] \\ &\leq g(c)P[|X| \leq c] + MP[|X| > c] \\ &\leq g(c) + MP[|X| > c]. \end{aligned}$$

277. Since $P[g(|X|) \leq g(M)] = 1$, then from Exercise 276 we have

$$E[g(X)||X| > c] \leq g(M).$$

278.
$$\begin{aligned} P\left[\max\left(\frac{|X_1 - \mu_1|}{\sigma_1}, \frac{|X_2 - \mu_2|}{\sigma_2}\right) < c\right] \\ = P\left[\frac{|X_1 - \mu_1|}{\sigma_1} < c, \frac{|X_2 - \mu_2|}{\sigma_2} < c\right] \\ = P[(x_1, x_2) \in R], \end{aligned}$$

where R is the rectangle defined by the lines $x_i = \mu_i \pm c\sigma_i$ $(i = 1, 2)$. The function

$$g(x_1, x_2, t) = \frac{1}{c^2(1 - t^2)}\left[\left(\frac{X_1 - \mu_1}{\sigma_1}\right)^2 - 2t\left(\frac{X_1 - \mu_1}{\sigma_1}\right)\left(\frac{X_2 - \mu_2}{\sigma_2}\right)\right.$$
$$\left. + \left(\frac{X_2 - \mu_2}{\sigma_2}\right)^2\right] \qquad \text{for} \quad |t| < 1,$$

is nonnegative and larger than 1 for $(x_1, x_2) \notin R$. We then have,

$$P[(X_1, X_2) \in R] = \int\int_R f(x_1, x_2)\, dx_1\, dx_2$$

$$= 1 - \int_{-\infty}^{\infty}\int_{-\infty}^{\infty} g(x_1, x_2, t) f(x_1, x_2)\, dx_1\, dx_2$$

$$= 1 - \frac{2(1 - t\rho)}{c^2(1 - t^2)}, \qquad |t| < 1.$$

The expression $[2(1 - t\rho)]/[c^2(1 - t^2)]$ is minimized when

$$t = \frac{1}{\rho}(1 - \sqrt{1 - \rho^2})$$

and thus

$$P\left[\frac{|X_1 - \mu_1|}{\sigma_1} < c, \frac{|X_2 - \mu_2|}{\sigma_2} < c\right] \geq 1 - \frac{1 - \sqrt{1 - \rho^2}}{c^2},$$

from which the inequality follows.

279. For every vector t, the function

$$g_t(X) = t'G(X)t$$

is a convex scalar function of the matrix X and by Jensen's inequality we have

$$g_t(EX) \leq E[g_t(X)],$$

that is, for every t,

$$t'G(EX)t' \leq Et'G(X)t.$$

Hence the relation

$$G(EX) \leq E[G(X)]$$

follows.

280. By Exercise 279, it suffices to show that the matrix function $G(X) = X^{-1}$ is convex. For this it must be shown that for symmetric matrices $X > 0$, $Y > 0$, of order r (say) and $0 \leq \lambda \leq 1$, the relation

$$[\lambda X + (1 - \lambda)Y]^{-1} \leq \lambda X^{-1} + (1 - \lambda)Y^{-1} \tag{1}$$

holds. There is a nonsingular matrix A such that $X = AA'$, $Y = ADA'$, where D a diagonal matrix with elements $d_1, \ldots, d_r$. Hence (1) is equivalent to

$$[\lambda I + (1 - \lambda)D]^{-1} \leq \lambda I + (1 - \lambda)D^{-1},$$

which is satisfied by diagonal matrices; the assertion now follows from the convexity of the scalar function $y = x^{-1}$.

Note. For $0 < \lambda < 1$, (1) holds with strict inequality.

281. From $(d^2/dy^2) \log y = -1/y^2 < 0$ it follows that $g(y) = \log y$ is a concave function; therefore, by Jensen's inequality,

$$E[\log Y] \le \log E(Y).$$

282. If we consider the random variable $Y_n = \sum_{i=1}^{n} \log X_i$ where X_1, $X_2, \ldots, X_n$ are independent and isonomic, it follows that $E(Y_n) = nE[\log X_i]$ and $\text{Var}(Y_n) = n \, \text{Var}(\log X_i)$. By Chebyshev's inequality, for every $\varepsilon > 0$,

$$P[|Y_n - E(Y_n)| < n\varepsilon] \ge 1 - \frac{\text{Var}(Y_n)}{n^2 \varepsilon^2},$$

or

$$P[n(E(\log X_i) - \varepsilon)] < \sum_{i=1}^{n} \log X_i < n[E(\log X_i) + \varepsilon]] \ge 1 - \frac{\sigma^2}{n\varepsilon^2},$$

or

$$P[\exp\{n(E(\log X_i) - \varepsilon)\}] < X_1 X_2 \ldots X_n < \exp\{n(E(\log X_i) + \varepsilon)\}$$

$$\ge 1 - \frac{\sigma^2}{n\varepsilon^2}.$$

Hence, and since by Exercise 281 $E[\log X] \le \log E(X)$, we obtain the required one-sided inequality.

283. The area of the triangle is $E = \frac{1}{2}XY$, where X and Y are independent and uniform in the intervals $(0, a)$ and $(0, b)$, respectively. (a and b denote the length and width of the rectangle, respectively). The density of E is

$$f(x) = \frac{2}{ab} \log(ab) + \frac{2}{ab} \log 2x, \qquad 0 < x < \tfrac{1}{2}ab.$$

Hence calculate the required probability.

284. Let A be the origin of the axes and α the abscissa of the point B. Then the randomly selected points X_1 and X_2 have the uniform distribution in $(0, \alpha)$ and the density of the ordered sample $X < Y$ (say) is

$$f(x, y) = \frac{2}{\alpha^2}, \qquad 0 < x < y < \alpha.$$

In order that X, Y be such that the segments $(AX), (XY), (YB)$ form a triangle, they must satisfy

$$2X < \alpha, \qquad Y - X < \frac{\alpha}{2} \quad \text{and} \quad Y > \frac{\alpha}{2};$$

these define the triangle (CDE) (see figure) Hence

$$P[(X, Y) \in (CDE)] = \frac{2}{\alpha^2} \times \text{area of } (CDE) = \frac{1}{4}.$$

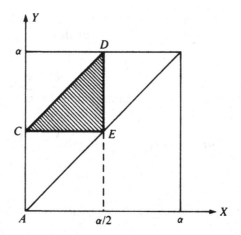

285. The first point P_1 can be selected as any point of the circumference and may be regarded as the origin of measuring the angles which define the positions of the other points on the circumference (see figure). For every θ, the probability (elementary) that θ_2, which defines the position of P_2, is in the interval $(0, \theta)$, and that θ_3 (for the third point P_3) is in the interval $(\theta, \theta + d\theta)$, is equal to

$$\frac{\theta}{2\pi} \frac{d\theta}{2\pi} = \frac{1}{4\pi^2} \theta \, d\theta.$$

Hence the required probability, say p_3, will be

$$p_3 = 3! \int_0^\pi \frac{1}{4\pi^2} \theta \, d\theta = \frac{3}{4\pi^2} \theta^2 \bigg|_0^\pi = \frac{3}{4}.$$

The factor 3! is needed because there are 3! different permutations of the three points P_1, P_2, P_3.

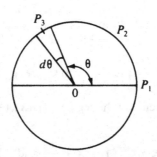

286. Let p be the required probability and let (AB) be the given line segment. Then

$$1 - p = P\left[\bigcup_{i=1}^5 A_i\right], \tag{1}$$

where A_i is the event that the ith part is larger than 1/2. Since evidently $A_i A_j = \varnothing$ $(i \neq j)$, we obtain from (1)

$$p = 1 - \sum_{i=1}^{5} P(A_i) = 1 - 5\left(\frac{1}{2}\right)^4,$$

because the probability that any part is larger than 1/2 is equal to the probability that each of the four points is contained in a subset of AB with length 1/2, that is,

$$P(A_i) = \left(\frac{1}{2}\right)^4.$$

287. As in Exercise 284, we find that the points X, Y must satisfy the relations

$$X < \frac{a+b}{2}, \qquad X + Y > \frac{a+b}{2} \qquad \text{and} \qquad Y < \frac{a+b}{2},$$

which define the triangle $\Theta Z H$ (shaded area). Hence

$$P[(X, Y) \in (\Theta H Z)] = \frac{\text{area of } (ZH\Theta)}{\text{area of } (O\Gamma E\Delta)} = \frac{b}{2a}.$$

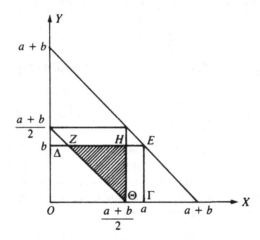

288. Let X be the distance of the nearest line to the middle M of the needle AB and φ the formed angle. A necessary and sufficient condition for the needle to cross the line is

$$X \leq \mu \sin \varphi$$

and the required probability p, in view of the uniform distribution of X in

$(0, a)$ and φ in $(0, \pi)$, is

$$p = \frac{1}{a\pi} \int_0^\pi \mu \sin \varphi \, d\varphi = \frac{2\mu}{a\pi}.$$

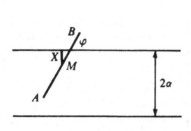
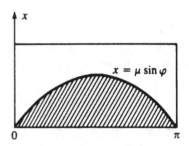

289. At first we consider a convex polygon $A_1 A_2 \cdots A_n$ (with diameter less than $2a$). Then a given line (of the parallel ones) can be crossed either by none of the sides of the polygon or by two sides only. The required probability p is given by

$$p = \sum_{i<j} p_{ij}, \qquad i, j = 1, 2, \ldots, n, \tag{1}$$

where $p_{ij} = P$ (of crossing sides i and j). On the other hand, the probability p_i (say) of being crossed by the side i, by Exercise 288, is given by

$$p_i = \sum_{i \neq j} p_{ij} = \frac{\mu_i}{\pi a}, \tag{2}$$

where μ_i the length of side i $(i = 1, 2, \ldots, n)$. The probability p, by (1) and (2), is

$$p = \sum_{i<j} p_{ij} = \frac{1}{2} \sum_{i \neq j} p_{ij} = \frac{1}{2} \sum_{i=1}^n p_i = \frac{\sum_{i=1}^n \mu_i}{2\pi a} = \frac{P}{2\pi a}, \tag{3}$$

where P is the perimeter of the polygon. Observing that p is independent of the number n of the sides of the convex polygon and depends only on its perimeter, going to the limit when $n \to \infty$ and max $\mu_i \to 0$, we see that the required probability for a convex and closed curve is given by (3), where P is the perimeter (length) of the curve.

290. Let a_n and b_n be the expected numbers of 10-drachma coins of A and B, respectively, after n exchanges. Then

$$a_n + b_n = a + b \qquad \text{and} \qquad a_0 = a. \tag{1}$$

Moreover, the difference equation

$$a_n = a_{n-1} + 1 \cdot P[E_n] + (-1)P[E_n^c] \tag{2}$$

holds, where E_n is the event that A gets a 10-drachma coin the nth exchange.

Since

$$P[E_n] = \frac{b_{n-1}}{b} = 1 - P[E_n^c],$$

relation (2) becomes

$$a_n = a_{n-1} + \frac{b_{n-1}}{b} - \frac{a_{n-1}}{a}, \qquad n = 1, 2, \ldots .$$

Taking into account (1), we obtain its general solution

$$a_n = \frac{a^2}{a+b} + \frac{ab}{a+b}\left(1 - \frac{1}{a} - \frac{1}{b}\right)^n, \qquad n = 1, 2, \ldots .$$

Hence the required probability p_n (say) that A gives a 10-drachma coin to B after n exchanges is

$$p_n = \frac{a_n}{a} = \frac{a}{a+b} + \frac{b}{a+b}\left(1 - \frac{1}{a} - \frac{1}{b}\right)^n.$$

291. (a) The required probability p_n (say) that the game finishes with an ace after no more than n tosses is

$$p_n = p_a^2 + p_a^2 \sum_{k=3}^{n} \binom{k-1}{1} p_0^{k-2} + p_a^2 p_b \sum_{k=3}^{n} (k-1)(k-2)p_0^{k-3}$$

$$= \frac{p_a^2(1 - p_0 + 2p_b)}{(1 - p_0)^3}[1 - np_0^{n-1} + (n-1)p_0^n] - \frac{(n-1)p_a^2 p_b p_0^{n-2}}{1 - p_0}, \quad (1)$$

where

$$p_a = P(\text{of ace}) = 1/6, \qquad p_b = P(\text{of six}) = 1/6,$$

and

$$p_0 = P(\text{neither ace nor six}) = 2/3.$$

Relation (1) is found from the fact that the event A_k that the game finishes after k tosses with an ace occurs if: (i) One ace appears on the first $k-1$ tosses, neither ace nor six on $k-2$ tosses, and ace on the kth toss. (ii) One ace and one six appear on the first $k-1$ tosses (hence neither ace nor six during $k-3$ tosses), and a six appears on the kth toss.

(b) The probability, p (say), that the game finishes with ace is the limit of p_n as $n \to \infty$. Thus from (1) we find

$$p = \frac{p_a^2(1 - p_0 + 2p_b)}{(1 - p_0)^3} = \frac{1}{2}.$$

Similarly, the probability that the game terminates with six is $1/2$ and therefore the game terminates with probability 1. For fixed n there is a positive probability $1 - 2p_n$ that the game continues after n tosses.

292. We have

$$p_n = (1 - p_{n-1})P(\text{of selection of } A \text{ in the } n\text{th year}) = \tfrac{1}{2}(1 - p_{n-1})$$

with $p_1 = 1$. The general solution of the difference equation

$$p_n + \tfrac{1}{2}p_{n-1} = \tfrac{1}{2}$$

is

$$p_n = \tfrac{1}{3}[1 + (-1)^{n-1}(\tfrac{1}{2})^{n-2}], \qquad n = 1, 2, \ldots.$$

293. p_n satisfies the difference equation of order 2

$$p_n - p_{n-1} + pqp_{n-2} = 0, \quad n \geq 2, \qquad p_0 = p_1 = 1.$$

(i) For $p \neq q$ it has the solution

$$p_n = \frac{p^{n+1} - q^{n+1}}{p - q}, \qquad n \geq 0.$$

(ii) For $p = q = 1/2$ we have

$$p_n = (n + 1)2^{-n}, \qquad n \geq 0.$$

294. (a) We have

$$y_{k+1} = y_k + 1 \cdot P(\text{the } k + 1 \text{ ball is placed in an empty cell})$$

$$= y_k + \frac{n - y_k}{n} = \left(1 - \frac{1}{n}\right)y_k + 1.$$

The resulting difference equation

$$y_{k+1} - \left(1 - \frac{1}{n}\right)y_k = 1$$

has the general solution

$$y_k = n + c\left(1 - \frac{1}{n}\right)^k,$$

where $c = -n$ since $y_1 = 1$.

(b) $\quad y_k = E\left[\sum_{i=1}^{n} X_i\right], \qquad$ where $\quad X_i = \begin{cases} 1 & \text{if cell } i \text{ is occupied,} \\ 0 & \text{if cell } i \text{ is empty.} \end{cases}$

Then

$$E(X_i) = 1 - \left(1 - \frac{1}{n}\right)^k, \qquad i = 1, 2, \ldots, n,$$

and thus we find

$$y_k = n\left[1 - \left(1 - \frac{1}{n}\right)^k\right].$$

295. Let A be the event of a double six in a toss of two dice. Then we have

$$p_n = p_{n-1} \cdot P[A^c] + (1 - p_{n-1}) \cdot P[A] = p_{n-1} \cdot \frac{35}{36} + (1 - p_{n-1})\frac{1}{36}.$$

Hence the required difference equation, with particular solution $p_n = a = 1/2$, and general solution of the corresponding homogeneous $p_n = c \cdot (17/18)^n$. Thus the general solution is $p_n = 1/2 + c(17/18)^n$, where from the initial condition $p_0 = 1$ we find that $c = 1/2$. Hence

$$p_n = \frac{1}{2}\left[1 + \left(\frac{17}{18}\right)^n\right], \qquad n = 0, 1, 2, \ldots.$$

296. Let p_k be the probability that a white ball is drawn bearing the number k $(k = 1, 2, \ldots, n)$. Then p_n satisfies the relation

$$p_n = \frac{\Lambda + 1}{\Lambda + M + 1}p_{n-1} + \frac{\Lambda}{\Lambda + M + 1}(1 - p_{n-1})$$

$$= \frac{1}{\Lambda + M + 1}p_{n-1} + \frac{\Lambda}{\Lambda + M + 1}$$

with initial condition

$$p_1 = \frac{\Lambda}{\Lambda + M}. \tag{1}$$

We find

$$p_n = \frac{\Lambda}{\Lambda + M} + c\frac{1}{(\Lambda + M + 1)^n},$$

where, in view of (1), $c = 0$. Hence

$$p_n = \frac{\Lambda}{\Lambda + M}, \qquad n = 1, 2, \ldots.$$

Remark. If we suppose that the ball transferred first is white, then

$$p_1 = 1 \quad \text{and} \quad p_n = \frac{\Lambda}{\Lambda + M} + \frac{M}{(\Lambda + M)(\Lambda + M + 1)^{n-1}} \xrightarrow[n \to \infty]{} \frac{\Lambda}{\Lambda + M}.$$

297. Let p_n be the required probability. Then it will satisfy the relation

$$p_n = pp_{n-1} + p'(1 - p_{n-1}), \qquad p_1 = p \qquad (n = 2, 3, \ldots).$$

Its general solution is

$$p_n = \frac{p'}{q + p'} + \frac{q}{q + p'}(p - p')^n.$$

For $q = 1 - p = p'$ it gives

$$p_n = \frac{1}{2} + \frac{1}{2}(p - q)^n \to \frac{1}{2} \qquad \text{as} \quad n \to \infty.$$

298. We have

$$p_n = p_{n-1}P(\text{of one ace}) + p_{n-2}P(\text{of double ace})$$

$$= p_{n-1}\frac{10}{36} + p_{n-2}\frac{1}{36},$$

with initial conditions $p_0 = 25/36$, $p_1 = 10/36 \times 25/360$. Solving this homogeneous difference equation we find p_n. For the probability, say q_n, that one obtains n points, when the game is not interrupted, we have the difference equation

$$q_n = \frac{25}{36} q_n + \frac{10}{36} q_{n-1} + \frac{1}{36} q_{n-2}.$$

Compare p_n and q_n.

299. Let Λ_n be the expected number of good tubes obtained after n trials and p_n the required probability. Then

$$\Lambda_{n+1} = \Lambda_n + 1 \cdot P(\text{of drawing a bad tube during the } (n+1) \text{ draw})$$

$$= \Lambda_{n+1}\left(1 - \frac{\Lambda_n}{a+b}\right)$$

and $\Lambda_0 = a$. Hence we find

$$\Lambda_n = (a+b) - b\left(1 - \frac{1}{a+b}\right)^n \quad \text{and} \quad p_n = \frac{\Lambda_n}{a+b}.$$

300. Define success (S) on the rth draw of a ball to be that the $(r-1)$ ball was drawn from the same cell $r = 2, \ldots, k$. The probability of such a success S is $1/n$. If in this sequence there is no sample SS (event A), then in the sequence of trials there is no cell receiving more than two balls successively. If q_k is the probability of A in a sequence $(S$ and $S^c)$ of n trials, then $p_k = q_{k-1}$ $(k > 2)$ and q_k satisfies the relation

$$q_k = \frac{n-1}{n} q_{k-1} + \frac{1}{n}\frac{n-1}{n} q_{k-2},$$

from which we have

$$p_k = p_k(n) = \frac{n-1}{n} p_{k-1} + \frac{n-1}{n} q_{k-2}, \qquad k > 2, \tag{1}$$

with $p_1 = q_0 = 1$, $p_2 = q_1 = 1$. If λ_1, λ_2 are the roots of the characteristic equation of (1), then p_n is given by

$$p_k = [(1 - \lambda_2)\lambda_1^k - (1 - \lambda_1)\lambda_2^{k-1}]/(\lambda_1 - \lambda_2).$$

(a) $\lim p_k = 0$ as $k \to \infty$, because $\lambda_1^k \to 0$, $\lambda_2^k \to 0$.
(b) $\lim p_k(n) = 1$ as $n \to \infty$, because $\lambda_1 \to 1$, $\lambda_2 \to 0$.

301. Applying Hölder's inequality (9.5) with $p = q = 1/2$ for the random variables $|X|^{(t+h)/2}$ and $|X|^{(t-h)/2}$ we have

$$E^2(|X|^t) \le E(|X|^{t+h})E(|X|^{t-h}), \qquad 0 \le h \le t. \tag{1}$$

Setting $t_1 = t + h$, $t_2 = t - h$, equation (1) becomes

$$E^2[|X|^{(t_1+t_2)/2}] \le E(|X|^{t_1})E(|X|^{t_2}),$$

or, taking logarithms,

$$2 \log E[|X|^{(t_1+t_2)/2}] \le \log E(|X|^{t_1}) + \log E(|X|^{t_2}).$$

Setting $g(t) = \log E(|X|^t)$, we have

$$g\left(\frac{t_1 + t_2}{2}\right) \le \tfrac{1}{2}g(t_1) + \tfrac{1}{2}g(t_2),$$

that is, $g(t)$ satisfies (9.1) with $\lambda = 1/2$ and hence it is convex since it is also continuous. Since $g(0) = 0$, the slope $g(t)/t$ of the line that connects the origin with the point $(t, g(t))$ increases as t increases and hence

$$e^{g(t)/t} = [E(|X|^t)]^{1/t}$$

is an increasing function of t.

302. $\varphi_0(t)$, as the real characteristic function of a random variable X, is given by

$$\varphi_0(t) = \int_{-\infty}^{\infty} \cos tx f_X(x)\, dx$$

and the required inequality for φ_0 follows by taking the expected value in the inequality

$$1 - \cos 2tX = 2(1 - \cos^2 tX) \le 4(1 - \cos tX).$$

For the other inequality we observe that if $\varphi(t)$ is the characteristic function of the random variable X then $|\varphi(t)|^2 = \varphi(t)\overline{\varphi(t)}$ is the characteristic function of $X_1 - X_2$, where X_1, X_2 are independent having the same distribution as X, and it is a real function. Taking $\varphi_0(t) = |\varphi(t)|^2$ in the above inequality, we have the required inequality.

303. Applying the Schwartz inequality

$$\left(\int fg\, dx\right)^2 \le \left(\int f^2\, dx \cdot \int q^2\, dx\right)$$

to

$$\varphi(t) = \int_{-\infty}^{\infty} (\cos tx) f(x)\, dx$$

with

$$f(x) = \cos tx \sqrt{f(x)}, \qquad g(x) = \sqrt{f(x)},$$

we have

$$|\varphi_0(t)|^2 \le \int_{-\infty}^{\infty} \cos^2 tx f(x)\, dx = \int_{-\infty}^{\infty} \frac{1}{2}(1 + \cos 2tx) f(x)\, dx.$$

Hence we have the required inequality.

304. $E(X_iX_j) = \text{Cov}(X_i, X_j)$ $(i \neq j)$, implies that at most one of the X_i, say X_1, has mean $\neq 0$. Since the central moments of order 3 are

$$E[(X_i - E(X_i))(X_j - E(X_j))(X_k - E(X_k))] = 0,$$

we have $E(X_2X_3X_4) = 0$. It must be shown now that for jointly normal random variables X_1, X_2, X_3, X_4 with mean 0, we have

$$\mu_{1234} = E(X_1X_2X_3X_4) = \sigma_{12}\sigma_{34} + \sigma_{14}\sigma_{23} + \sigma_{13}\sigma_{24}.$$

From the joint characteristic function of X_1, X_2, X_3, X_4

$$\varphi(t_1, t_2, t_3, t_4) = \exp\left[-\frac{1}{2}\sum_{i=1}^{4}\sum_{j=1}^{4}\sigma_{ij}t_it_j\right]$$

and from the relation (cf. 6.15)

$$\mu_{1234} = \frac{\partial^4\varphi(t_1, t_2, t_3, t_4)}{\partial t_1\,\partial t_2\,\partial t_3\,\partial t_4}\bigg|_{(0,0,0,0)},$$

(1) is readily obtained.

305. By Exercise 241, the density of $nX_{(1)}$ is

$$f_n(x) = \left(1 - \frac{x}{n}\right)^{n-1}, \qquad 0 < x < n,$$

and its characteristic function is

$$\varphi_n(t) = \int_0^n e^{itx}\left(1 - \frac{x}{n}\right)^{n-1}dx \xrightarrow[n\to\infty]{} \int_0^\infty e^{itx}e^{-x}\,dx,$$

that is, the characteristic function of an exponential with density $f(x) = e^{-x}$.

306. We have

$$E(\bar{Z}) = E(Z_j) = P[(X_j - \bar{X})(Y_j - \bar{Y}) > 0]$$

and let $\rho(X_j - \bar{X}, Y_j - \bar{Y}) = \rho^*$. Then (see Exercise 220),

$$E(Z_j) = \frac{1}{2} + \frac{\arcsin\rho^*}{\pi}.$$

307. Let $X_1, X_2, \ldots, X_r$ be the selected lucky numbers and

$$S_r = \sum_{i=1}^{r} X_i.$$

The X_i are isonomic random variables with distribution

$$P[X_i = k] = \frac{1}{n}, \qquad i = 1, 2, \ldots, r, \quad k = 1, 2, \ldots, n.$$

Hence we have

$$E(X_i) = \frac{1}{n} \sum_{k=1}^{n} k = \frac{n+1}{2},$$

$$E(X_i^2) = \sum_{k=1}^{n} \frac{1}{n} k^2 = \frac{1}{n}(1^2 + 2^2 + \cdots + n^2) = \frac{(n+1)(2n+1)}{6},$$

$$\sigma^2 = \text{Var}(X_i) = E(X_i^2) - [E(X_i)]^2 = \frac{n^2 - 1}{12}. \tag{1}$$

Therefore we have

$$E(S_r) = \frac{r(n+1)}{2},$$

$$\text{Var}(S_r) = \sum_{i=1}^{r} \text{Var}(X_i) + r(r-1) \text{Cov}(X_i, X_j) = r\sigma^2 + r(r-1)\rho\sigma^2, \tag{2}$$

where $\rho = \text{Cov}(X_i, X_j) = \rho(X_i, X_j)$, the correlation of X_i, Y_j. Since the joint distribution of X_i, Y_j is independent of r, taking $r = n$, relation (2) gives

$$\text{Var}(S_n) = \text{Var}\left(\sum_{i=1}^{n} X_i\right) = n\sigma^2 + n(n-1)\rho\sigma^2 = 0, \tag{3}$$

because $\sum_{i=1}^{n} X_i = \sum_{i=1}^{n} i = \text{constant}$. Hence $\rho = -1/(n-1)$ and from (1), (2), and (3) we find that

$$\text{Var}(S_r) = \frac{r(n^2 - 1)}{12}\left(1 - \frac{r-1}{n-1}\right).$$

308. The probability generating function of X, because of the independence of the X_i, is

$$P_n(t) = \prod_{i=1}^{n} P_{X_i}(t) = \prod_{i=1}^{n} (p_i t + q_i). \tag{1}$$

Taking logarithms in (1), we obtain

$$\log P_n(t) = \sum_{i=1}^{n} \log(p_i t + q_i) = \sum_{i=1}^{n} \log(1 + p_i(t-1))$$

$$= (t-1) \sum_{i=1}^{n} p_i - \frac{(t-1)^2}{2} \sum_{i=1}^{n} p_i^2 + \cdots. \tag{2}$$

By hypothesis

$$\sum_{i=1}^{n} p_i \to \lambda \qquad \text{as} \quad n \to \infty$$

and the right-hand member of (2) tends to $(t-1)\lambda$, that is, $P_n(t)$ tends to the probability generating function of the required Poisson.

309. Putting $q = \lambda/n$, we have

$$P[X = k] = \binom{n + k - 1}{k} q^k p^n = \prod_{i=0}^{k-1} \left(1 + \frac{i}{n}\right) \frac{\lambda^k}{k!} \left(1 - \frac{\lambda}{n}\right)^n,$$

and since

$$\left(1 + \frac{i}{n}\right) \to 1, \quad \left(1 - \frac{\lambda}{n}\right)^n \to e^{-\lambda} \quad \text{as} \quad n \to \infty,$$

it follows that

$$P[X = k] \to e^{-\lambda} \frac{\lambda^k}{k!}, \qquad k = 0, 1, 2, \dots.$$

310. X will have a Gamma distribution and the unconditional distribution of Y is a continuous mixture of the Poisson with parameter λ, which follows the Gamma distribution of X (see Exercise 327).

311. The distribution of $|X - \mu|/\sigma$ is exponential with density

$$f(x) = e^{-x}$$

and characteristic function (Exercise 202)

$$\varphi(t) = (1 - it)^{-1}.$$

Hence

$$\frac{n}{\sigma} Y = \sum_{j=1}^{n} \frac{|X_j - \mu|}{\sigma}$$

has characteristic function

$$\varphi_n(t) = \varphi^n(t) = (1 - it)^{-n}$$

and $(2n/\sigma) Y$ has characteristic function $(1 - 2it)^{-n}$, that is, of a χ^2_{2n}. Finally, from $E(\chi^2_{2n}) = 2n$, it follows that $E(Y) = \sigma$. Thus the so-called mean deviation of the sample, that is, Y in the case of a Laplace distribution, gives an unbiased estimator of the standard deviation σ of the distribution.

312. $\sum \frac{X_i^2}{n} \xrightarrow{P} E(X_i^2) = 1$ (WLLN)   and $S_n \sqrt{n} \xrightarrow{L} N(0, 1)$ (CLT).

According to Exercise 255(c),

$$Y_n \xrightarrow{L} N(0, 1).$$

Similarly for Z_n.

313. The distribution of $X_{(1)}, \dots, X_{(r)}$, given that $X_{(r+1)} = x$, is the distribution of the ordered sample of r independent uniform $X_1, \dots, X_r$ in $(0, x)$. Moreover, $X_1 + \cdots + X_r = X_{(1)} + \cdots + X_{(r)} = Y_r$ and hence the conditional distribution of Y_r, given that $X_{(r+1)} = x$, is the distribution of the sum of observations of a random sample from the uniform in $(0, x)$. Thus, the conditional distribution of $Y_r/X_{(r+1)}$, given $X_{(r+1)} = x$, is the distribution of the sum of r independent uniform variables $X_1/x, \ldots, X_r/x$ in $(0, 1)$.

314. By induction. For $n = 2$ we have, by the convolution formula,

$$f_2(x) = \begin{cases} \displaystyle\int_0^x dy = x, & x \le 1 \\[2mm] \displaystyle\int_0^{1-x} dy = 2 - x, & 1 \le x \le 2 \end{cases}$$

$$= \frac{1}{(2-1)!} \left\{ \binom{2}{0} x - \binom{2}{1}(x-1) \right\}.$$

Suppose the relation holds for $n = k$, that is,

$$f_k(x) = \frac{1}{(k-1)!} \left\{ \binom{k}{0} x^{k-1} - \binom{k}{1}(x-1)^{k-1} + \cdots \right\}.$$

We shall show that it holds for $n = k + 1$. Indeed, we have

$$f_{k+1}(x) = \int_{x-1}^x f_k(y) f_1(x - y)\, dy.$$

For $0 < x < 1$ the lower limit of integration is 0, while for $k \le x \le k + 1$ this is equal to $x - 1$; the upper limit is x for $f_k(y) > 0$ and $0 < y \le k$. Hence the assertion follows.

The second expression of $f_n(x)$ follows from the first one if x is replaced by $n - x$, since if $X_1, \ldots, X_n$ are uniform in $(0, 1)$, then $1 - X_1, \ldots, 1 - X_n$ are also uniform in $(0, 1)$ and therefore the distribution of S_n is the same as that of $n - S_n$.

315. The probability generating function of S_n is $P(t) = P^n_{X_i}(t)$ where

$$P(t) = \left[\frac{1 - t^m}{m(1 - t)} \right]^n = \sum_{k=0}^{\infty} t^k P[S_n = k].$$

From this we find

$$P[S_n = k] = \frac{1}{m^n} \sum_{i=0}^{\infty} (-1)^{k+i+im} \binom{n}{i} \binom{-n}{k - mi}, \tag{1}$$

where, in reality, the sum extends only over all i for which the relation $k - mi \ge 0$ holds.

316. Because of (1) of Exercise 315, the required probability is

$$P[S_n = k - n] = P[S_3 = 10 - 3] = P[S_3 = 7]$$

$$= \frac{1}{6^3} \left[(-1)^7 \binom{3}{0} \binom{-3}{7} + (-1)^{14} \binom{3}{1} \binom{-3}{1} \right]$$

$$= \frac{1}{6^3} (36 - 9) = \frac{1}{8}.$$

317. Let $C_1, \ldots, C_{m+1}$ denote the $r = m + 1$ boxes. Then

(a) $P[X_1 = x_1, \ldots, X_m = x_m] = \sum_{i=1}^{r} P[A_i, X_1 = x_1, \ldots, X_m = x_m]$,   (1)

where, e.g.,

$$P[A_r, X_1 = x_1, \ldots, X_m = x_m] = \frac{\Gamma(rN - x_0)}{\Gamma(N + 1) \prod_{i=1}^{m} (N - x_i)!} r^{x_0 - rN - 1}, \quad (2)$$

where we put $x_0 = x_1 + \cdots + x_m$, since C_r is found empty (event A_r) and $x_1, \ldots, x_m$ matches in the remaining m boxes requires that $N - x_i$ matches are taken from C_i $(i = 1, 2, \ldots, m)$ before the $(N + 1)$st selection of C_i $(i = 1, 2, \ldots, m)$ (negative multinomial, see (s.7)) and

$$P \text{ (of selection of } C_i \text{ in every draw)} = \frac{1}{r}, \quad i = 1, 2, \ldots, r.$$

But, by symmetry, the r probabilities in (1) are equal and therefore

$$P[X_1 = x_1, \ldots, X_m = x_m]$$
$$= rP[A_k, X_1 = x_1, \ldots, X_m = x_m]$$
$$= \frac{\Gamma(rN - x_0)}{\Gamma(N + 1) \prod_{i=1}^{m} (N - x_i)!} r^{x_0 - rN}, \quad x_i = 0, 1, \ldots, N.$$

(b) Now $N - x_i$ matches must be drawn from C_i before the Nth selection of C_r. Thus

$$P[X_1 = x_1, \ldots, X_m = x_m] = \frac{\Gamma(rN - x_0)}{\Gamma(N) \prod_{i=1}^{m} (N - x_i)!} r^{x_0 - rN - 1}.$$

(Here $x_i = 1, 2, \ldots, N$ $(i = 1, 2, \ldots, m)$.)

318. The required probability, say a_r, is equal to

$$a_r = \sum_{x_1=1}^{N} \cdots \sum_{x_m=1}^{N} P[X_1 = x_1, \ldots, X_m = x_m],$$

where, putting $x_0 = x_1 + \cdots + x_m$, we have (cf. Exercise 317)

$$P[X_1 = x_1, \ldots, X_m = x_m] = \frac{\Gamma(rN - x_0)}{\Gamma(N)} \prod_{i=1}^{m} \frac{p_i^{N - x_i}}{(N - x_i)!}.$$

In the special case $r = 2$, a_2 can be written as an incomplete B function (see Exercise 329):

$$a_2 = \frac{\Gamma(2N)}{\Gamma(N)} \int_0^{p_2} x^{N-1}(1 - x)^{N-1} \, dx \equiv I_{p_2}(N, N).$$

319. By definition of independence the conditions are necessary. That they are sufficient follows from the fact that the second one gives

$$P[X_1 = x_1, X_2 = x_2] = P[X_1 = x_1] \cdot P[X_2 = x_2] \qquad \text{for all } (x_1, x_2). \quad (1)$$

The first one gives

$$P[X_3 = x_1, X_2 = x_2, X_1 = x_1] = \frac{P[X_1 = x_1, X_2 = x_2, X_3 = x_3]}{P[X_1 = x_1, X_2 = x_2]}$$

$$= P[X_3 = x_3].$$

This with (1) gives the condition of complete independence

$$P[X_1 = x_1, X_2 = x_2, X_3 = x_3] = P[X_1 = x_1]P[X_2 = x_2]P[X_3 = x_3]$$

for all (x_1, x_2, x_3). For n variables $X_1, \ldots, X_n$, the conditions become

$$P[X_i = x_i | X_j = x_j, j = 1, 2, \ldots, i-1] = P[X_i = x_i], \qquad i = 1, 2, \ldots, n.$$

320. $E(S_n^{-1})$ exists because $P[S_n^{-1} < X_1^{-1}] = 1$. Since $E(X_i/S_n) = a$ $(i = 1, 2, \ldots, n)$, we have

$$1 = E\left(\sum_{i=1}^{n} \frac{X_i}{S_n}\right) = \sum_{i=1}^{n} E\left(\frac{X_i}{S_n}\right) = nE\left(\frac{X_i}{S_n}\right).$$

Hence,

$$a = E\left(\frac{X_i}{S_n}\right) = n^{-1}.$$

321. By Exercise 320, we have

$$E\left(\frac{S_m}{S_n}\right) = mE\left(\frac{X_i}{S_n}\right) = \frac{m}{n} \qquad \text{for} \quad m \le n,$$

while, for $m \ge n$, since S_n^{-1} and $X_{n+1}, \ldots, X_m$ are independent, we have

$$E\left(\frac{S_m}{S_n}\right) = E\left(1 + \frac{1}{S_n}\sum_{i=n+1}^{m} X_i\right) = 1 + (m-n)E\left(\frac{X_i}{S_n}\right)$$

$$= 1 + (m-n)E(X_i)E\left(\frac{1}{S_n}\right).$$

322. (a) The required probability p is given by

$$p = \sum_{n=1}^{\infty} P(A_n) = \sum_{n=1}^{\infty} \frac{1}{3}\left(\frac{2}{3}\right)^{n-1} = 1,$$

where A_n is the event that it gets stuck exactly after n steps, i.e., at time $t = n$ (geometric distribution with $p = 1/3$).

(b) Let p_a, p_b, p_c be the probabilities that the caterpillar eventually gets stuck coming from A, B, C, respectively. Then we have

$$p_a + p_b + p_c = 1, \qquad p_b = p_c = \tfrac{1}{3}p_a,$$

since the caterpillar goes in one step to B with probability $1/3$ and having gone there it has the same probability of getting stuck (coming from B) as if it had started at A. Hence

$$p_b = p_c = 1/5, \qquad p_a = 3/5.$$

323. (a) (i) $E(X_n) = naE(\cos \theta_i) = na(2\pi)^{-1} \int_0^{2\pi} \cos \theta \, d\theta = 0.$

Similarly, $E(Y_n) = 0$,

$$\text{Var}(X_n) = E(X_n^2) = na^2 E(\cos^2 \theta_i) + n(n-1)a^2 E(\cos \theta_i \cos \theta_j)$$

$$= na^2 E(\cos^2 \theta_i), \tag{1}$$

since the independence of θ_i gives

$$E(\cos \theta_i \cos \theta_j) = E(\cos \theta_i)E(\cos \theta_j) = 0, \qquad i \neq j.$$

From (1), we find

$$\text{Var}(X_n) = \tfrac{1}{2}na^2. \qquad \text{Similarly,} \qquad \text{Var}(Y_n) = \tfrac{1}{2}na^2. \tag{2}$$

(ii) From (1) and (2) we find

$$E(R_n^2) = na^2.$$

(b) $\qquad \text{Cov}(X_n, Y_n) = E(X_n Y_n) = na^2 [E(\sin \theta_i \cos \theta_i)$

$$+ (n-1)E(\sin \theta_i \cos \theta_j)] = 0.$$

Hence the coordinates X_n, Y_n of the position of the flea after n jumps are uncorrelated. However, we shall show that they are not independent. Let $n = 1$. Then we have, e.g.,

$$P\left[X_1 > \frac{a}{\sqrt{2}}, Y_1 > \frac{a}{\sqrt{2}} \right] = 0 \neq P\left[X_1 > \frac{a}{\sqrt{2}} \right] P\left[Y_1 > \frac{a}{\sqrt{2}} \right],$$

that is, X_1, Y_1 are not independent. Similarly, by induction, the dependence of X_n, Y_n is shown.

(c) By the CLT we have

$$X_n \xrightarrow{L} N\left(0, \frac{na^2}{2} \right), \qquad Y_n \xrightarrow{L} N\left(0, \frac{na^2}{2} \right).$$

Hence X_n, Y_n are asymptotically normal and independent as well, because they are uncorrelated. Then, the asymptotic (as $n \to \infty$) distribution of R_n, by Exercise 221, has the density

$$f_{R_n}(r) = \frac{2r}{na^2} e^{-r^2/na^2}.$$

From the above relation we conclude that the expected distance of the flea from the origin (for large n) is

$$E(R_n) = \sqrt{n\pi}\,\frac{a}{2}.$$

324. Let $\xi_j = (X_1, \ldots, X_{j-1}, X_{j+1}, \ldots, X_n)$ $(j = 1, 2, \ldots, n)$. The density of ξ_j is

$$f_j(\xi_j) = \int_{-\infty}^{\infty} f(x)\, dx_j = I_1 + I_2,$$

where

$$I_1 = \int_{-\infty}^{\infty} \frac{1}{(2\pi)^{n/2}} \exp\left[-\frac{1}{2}\sum_{i=1}^{n} x_i^2\right] dx_j = \frac{1}{(2\pi)^{(n-1)/2}} \exp\left[-\frac{1}{2}\sum_{i \neq j} x_i^2\right]$$

is the joint density of $n - 1$ independent $N(0, 1)$, and

$$I_2 = \frac{1}{(2\pi)^{n/2}} \exp\left[-\frac{1}{2}\sum_{i=j} x_i^2\right]\left[\prod_{i \neq j} x_i e^{-x_i^2/2}\right] \cdot \int_{-\infty}^{\infty} x_j e^{-x_j^2/2}\, dx_j = 0.$$

Hence, any subvector of ξ_j and therefore any subset of $X_1, \ldots, X_n$ has the multinormal distribution with independent components. Yet, $X = (X_1, \ldots, X_n)$ with density $f(x)$ does not have an n-dimensional normal distribution and $X_1, X_2, \ldots, X_n$ are not completely independent.

Remark. If $X_1, \ldots, X_n$ are uncorrelated, that is, $\rho(X_i, X_j) = 0$ $(i \neq j)$, and all the marginal distributions are normal, then $X_1, X_2, \ldots, X_n$ are jointly normal if and only if $X_1, X_2, \ldots, X_n$ are completely independent.

325. (a) Because of the uniform distribution of the angle θ, Exercise 288, the mean number of crossings of the needle by horizontal or vertical lines is the same, and equal to the probability of crossing one of the parallels, i.e., $2\mu/\pi a$. But the mean number of crossings of the lines of the table is equal to the sum of the expected crossings of the horizontal and vertical lines, i.e., $4\mu/\pi a$. Thus for a needle having length equal to the side of the squares, the mean number of crossings is $4/\pi \approx 1.27$.

(b) If the needle has arbitrary length l, let us imagine that this is divided into n equal pieces of length less than $2a$. Each of the pieces separately gives mean number of crossings equal to $2l/n\pi a$. But the mean of the sum is equal to the sum of the means and hence the required mean number of crossings by the needle is $2l/\pi a$. Throwing the n pieces of the needle separately instead of throwing the needle as a solid body does not influence the mean.

Note. Use the result for the estimation of the value of π, e.g., throwing a toothpick on a flat surface divided into little squares.

326. (Cacoullos, *JASA*, 1965)
(a) We have

$$G(u) = P\left[\frac{X - Y}{2\sqrt{XY}} \leq u\right] = \frac{1}{2^n \Gamma^2(n/2)} \int_{(x-y)=2u\sqrt{xy}} (xy)^{(n/2)-1} e^{-(x+y)/2}\, dx\, dy,$$

where putting first $x = \rho \cos^2 \theta$, $y = \rho \sin^2 \theta$, and then $w = \cot 2\theta$, we have by the Legendre duplication formula,

$$2^{2n-1}\Gamma(n)\Gamma(n + \tfrac{1}{2}) = \sqrt{\pi}\,\Gamma(2n),$$

$$G(u) = \frac{\Gamma\left(\dfrac{n+1}{2}\right)}{\sqrt{\pi}\,\Gamma\left(\dfrac{n}{2}\right)} \int_{-\infty}^{u} \frac{dw}{(1 + w^2)^{(n+1)/2}}.$$

Hence we obtain the required distribution of Z. Since Z can be written as

$$Z = \frac{\sqrt{n}}{2}\left(\sqrt{F_{n,n}} - \frac{1}{\sqrt{F_{n,n}}}\right),$$

we have the relation between t_n and $F_{n,n}$.

(b) For $0 < \alpha < 0.5$ we have (for $F_{n,n} = 1$, $t_n = Z = 0$)

$$P[0 < t_n < t_n(\alpha)] = P[1 < F_{n,n} < F_\alpha], \tag{1}$$

where F_α is the solution $F > 1$ of the equation

$$t_n(\alpha) = \frac{\sqrt{n}}{2}\left(\sqrt{F} - \frac{1}{\sqrt{F}}\right), \tag{2}$$

that is,

$$F_a = 1 + \frac{2t_n^2(\alpha)}{n} + \frac{2t_n(\alpha)}{\sqrt{n}}\sqrt{1 + \frac{t_n^2(\alpha)}{n}}.$$

From (1) and from the relation $P[F_{n,n} \le 1] = 1/2$, we have

$$P[F_{n,n} \le F_\alpha] = 1 - \alpha$$

and hence $F_{n,n}(\alpha) = F_\alpha$. For $0.5 < \alpha < 1$, as is well known, we have

$$t_n(\alpha) = -t_n(1 - \alpha) \qquad \text{and} \qquad F_{n,n}(\alpha) = F_{n,n}^{-1}(1 - \alpha).$$

327. According to the generalized theorem of total probability, we have

$$p_k = \int_0^\infty e^{-\lambda}\frac{\lambda^k}{k!} f(\lambda)\,d\lambda = \frac{a^\rho}{\Gamma(\rho)k!}\int_0^\infty \lambda^{k+\rho-1}e^{-(a+1)\lambda}\,d\lambda$$

$$= \frac{a^\rho}{\Gamma(\rho)k!}\frac{\Gamma(k+\rho)}{(a+1)^{k+\rho}} = \binom{-\rho}{k}(-p)^k q^\rho,$$

where we made use of $\Gamma(x + 1) = x\Gamma(x)$ $(x > 0)$. Hence the negative binomial distribution was obtained as a mixture of the Poisson and Gamma distributions.

328. By definition we have

$$E[\text{Var}(X|Y)] = E[E(X^2|Y) - \{E(X|Y)\}^2]$$

$$= E[E(X^2|Y)] - E[E(X|Y)]^2, \tag{1}$$

$$\text{Var}[E(X|Y)] = E[E(X|Y)]^2 - \{E[E(X|Y)\}^2, \tag{2}$$

and since

$$E[E(X|Y)] = E(X), \qquad E[E(X^2|Y)] = E(X^2).$$

Adding (1) and (2), we have

$$E(X^2) - \{E(X)\}^2 = \text{Var}(X).$$

329. The distribution function of $X_{(k)}$, by Exercise 241, is

$$F_k(p) = P[X_{(k)} \leq p] = \int_0^p f_{X_k}(x)\, dx$$

$$= \frac{1}{B(k, n - k + 1)} \int_0^p x^{k-1}(1 - x)^{n-k}. \tag{1}$$

On the other hand, we have that $X_{(k)} \leq p$ if and only if at least k of the X_i are less than p. Since the X_i are independent and $P[X_i \leq p] = p$, we have

$$F_k(p) = \sum_{j=k}^{n} \binom{n}{j} p^j q^{n-j}. \tag{2}$$

From (1) and (2) the assertion follows.

Bibliography

1. Cacoullos, T. (1969). *A Course in Probability Theory* (in Greek). Leousis-Mastroyiannis, Athens.
2. Cramér, H. (1946). *Mathematical Methods of Statistics.* Princeton University Press, Princeton, NJ.
3. Dacunha-Castelle, D., Revuz, D., and Schreiber, M. (1970). *Recueil de Problemes de Calcul des Probabilités.* Masson Cie, Paris.
4. Feller, W. (1957). *Introduction to Probability Theory and its Applications*, Vol. I. Wiley, New York.
5. Fisz, M. (1963). *Probability Theory and Mathematical Statistics.* Wiley, New York.
6. Gnedenko, B.V. (1962). *The Theory of Probability.* Chelsea, New York.
7. Mosteller, F. (1965). *Fifty Challenging Problems in Probability.* Addison-Wesley, Boston, MA.
8. Parzen, E. (1960). *Modern Probability Theory and its Applications.* Wiley, New York.
9. Rahman, N.A. (1967). *Exercises in Probability and Statistics.* Griffin, London.
10. Sveshnikov, A.A. (1968). *Problems in Probability Theory.* Saunders, Philadelphia.

Index